Markus Bendele

The Superconducting and Magnetic Properties of the Iron-Chalcogenides

Markus Bendele

The Superconducting and Magnetic Properties of the Iron-Chalcogenides

Südwestdeutscher Verlag für Hochschulschriften

Impressum / Imprint
Bibliografische Information der Deutschen Nationalbibliothek: Die Deutsche Nationalbibliothek verzeichnet diese Publikation in der Deutschen Nationalbibliografie; detaillierte bibliografische Daten sind im Internet über http://dnb.d-nb.de abrufbar.
Alle in diesem Buch genannten Marken und Produktnamen unterliegen warenzeichen-, marken- oder patentrechtlichem Schutz bzw. sind Warenzeichen oder eingetragene Warenzeichen der jeweiligen Inhaber. Die Wiedergabe von Marken, Produktnamen, Gebrauchsnamen, Handelsnamen, Warenbezeichnungen u.s.w. in diesem Werk berechtigt auch ohne besondere Kennzeichnung nicht zu der Annahme, dass solche Namen im Sinne der Warenzeichen- und Markenschutzgesetzgebung als frei zu betrachten wären und daher von jedermann benutzt werden dürften.

Bibliographic information published by the Deutsche Nationalbibliothek: The Deutsche Nationalbibliothek lists this publication in the Deutsche Nationalbibliografie; detailed bibliographic data are available in the Internet at http://dnb.d-nb.de.
Any brand names and product names mentioned in this book are subject to trademark, brand or patent protection and are trademarks or registered trademarks of their respective holders. The use of brand names, product names, common names, trade names, product descriptions etc. even without a particular marking in this works is in no way to be construed to mean that such names may be regarded as unrestricted in respect of trademark and brand protection legislation and could thus be used by anyone.

Coverbild / Cover image: www.ingimage.com

Verlag / Publisher:
Südwestdeutscher Verlag für Hochschulschriften
ist ein Imprint der / is a trademark of
OmniScriptum GmbH & Co. KG
Heinrich-Böcking-Str. 6-8, 66121 Saarbrücken, Deutschland / Germany
Email: info@svh-verlag.de

Herstellung: siehe letzte Seite /
Printed at: see last page
ISBN: 978-3-8381-3768-1

Zugl. / Approved by: Zürich, Universität Zürich, Diss., 2011

Copyright © 2013 OmniScriptum GmbH & Co. KG
Alle Rechte vorbehalten. / All rights reserved. Saarbrücken 2013

Meinem Vater

Abstract

Superconductivity is a remarkable phenomenon that was discovered exactly 100 years ago by H. Kamerlingh-Onnes. Many famous physicists such as N. Bohr, W. Heisenberg, or A. Einstein, to name only a few, tried for more than 50 years to describe the mechanism that leads to superconductivity. Only in 1957 a theory was suggested by J. Bardeen, L. Cooper, and J. Schrieffer, that was widely accepted. 25 years ago the striking finding of high temperature superconductivity in copper based materials, the so-called cuprates, by J. G. Bednorz and K. A. Müller revolutionized the field of superconductivity. The superconducting transition temperature increased within a short period of time from 23 K in the materials known till then to approximately 140 K in the cuprates. Since the discovery a great effort has been made towards the understanding of the mechanism of high temperature superconductivity and the microscopic pairing mechanism. However, it remains one of the biggest mysteries in physics. Obviously the high temperature superconductors bear still lots of surprises as ten years ago the diborides were discovered to be superconducting and recently, only three years ago, the finding of the iron-based high temperature superconductors attracted again the attention to the field. To find the microscopic mechanism leading to superconductivity in the iron-based high temperature superconductors might help to resolve the mystery of high temperature superconductivity in general.

This book is focused on the simplest of the iron-based high temperature superconductors, namely the binary FeCh family. Here Ch stands for the chemical elements belonging to the chalcogenide group like Sulfur (S), Selenium (Se), and Tellurium (Te). It is the simplest among this class because of its simple crystallographic structure consisting of a stack of FeCh layers. Furthermore, it is an ideal modeling system for the other iron-based superconductors because of its simplicity and its similarity with their electronic structure.

The electronic phase diagrams of the FeCh family contain the appearance of different ground states. Whereas the mother compounds are in general antiferromagnetically ordered, the material becomes superconducting after going through a region where superconductivity and magnetism coexist. In the framework of this book, the FeCh system was tuned solely by changing the lattice either by hydrostatic or chemical pressure and without introducing additional charge carriers. The muon spin rotation/relaxation/resonance (μSR) technique in combination with ac and dc magnetization experiments is an ideal tool to investigate the superconducting and magnetic states and the interplay in a sense of competition and/or coexistence between them. It can be seen that the system is extremely sensitive to pressure. FeSe$_{1-x}$ at ambient pressure is superconducting and nonmagnetic. Upon applying hydrostatic pressure the superconducting transition temperature increases and exhibits one of the biggest pressure effects known. Surprisingly, the compound features the appearance of magnetism that coexists on atomic length scales with superconductivity at high pressures. A similar effect is observed if chemical pressure is applied by substituting Se by the isovalent Te. First the superconducting transition temperature increases, then the system becomes magnetic and superconducting at the same time, and finally it turns into an antiferromagnet. This suggests that the lattice plays a very important role in determining whether the system is superconducting, magnetic, or both at the same time as already minor changes lead to drastic differences in electronic characteristics.

In order to find hints for the pairing mechanism leading to superconductivity in the iron-based high temperature superconductors, the temperature dependence of the superfluid density and the resulting superconducting gap structure were examined. It is seen that all of the iron-based superconductors, including the FeCh, have multiple gaps that open below the critical temperature T_c. Interestingly, the gap to T_c ratio is similar in all the iron-based superconductors. A further important hint to the pairing mechanism may come from isotope exchange experiments. They in fact showed that the transition temperature is dependent on the isotope mass, which strongly suggests that the lattice effects play a major role in the pairing mechanism.

Zusammenfassung

Die Supraleitung ist ein bemerkenswertes Phänomen, das vor 100 Jahren vor H. Kamerlingh-Onnes entdeckt wurde. Daraufhin haben viele berühmte Physiker wie zum Beispiel N. Bohr, W. Heisenberg oder A. Einstein, versucht den Mechanismus, welcher zur Supraleitung führt, zu beschreiben. Dies gelang jedoch erst 1957 J. Bardeen, L. Cooper und J. Schrieffer. Vor 25 Jahren haben J. G. Bednorz und K. A. Müller das Feld der Supraleitung mit der Entdeckung der Hochtemperatursupraleitung revolutioniert. Lag die höchste supraleitende Sprungtemperatur bis dahin bei 23 K, erreichte sie danach innerhalb kurzer Zeit einen Wert von ca. 140 K. Seither sind grosse Anstrengungen unternommen worden, den Mechanismus der Hochtemperatursupraleitung und den mikroskopischen Paarungsmechanismus zu beschreiben. Dies ist jedoch bis heute nicht gelungen und die Supraleitung bleibt eines der grössten Rätsel der Physik, welches immer noch viele Überraschungen birgt. So wurde vor 10 Jahren entdeckt, dass die Diboride supraleitend sind und vor nur drei Jahren wurde die Entdeckung von Supraleitung in eisenbasierten Materialien gefeiert. Indem der Paarungsmechanismus, der zur Supraleitung führt, in den Letztgenannten gefunden wird, können Rückschlüsse auf alle Klassen gemacht werden und möglicherweise das Rätsel der Hochtemperatursupraleitung im Generellen gelöst werden.

In dieser Arbeit wird der Fokus auf das $FeCh$ System, die einfachste Familie der Eisenbasierten Hochtempratursupraleitern, gelegt. Hierbei steht Ch als Akronym für die chemischen Elemente welche zur Gruppe der Chalkogenide gehören. Das sind zum Beispiel Schwefel (S), Selen (Se) oder Tellur (Te). $FeCh$ wird aufgrund seiner einfachen Kristallstruktur, die nur aus einer Stapelung aus $FeCh$ Schichten besteht, als das einfachste System bezeichnet. Zusätzlich ist es ein ideales Modellsystem für die anderen Familien der Eisenbasierten Hochtemperatursupraleiter gerade wegen dieser Einfachheit und der sich extrem ähnlichen elektronischen Struktur zwischen allen Familien.

Die elektronischen Phasendiagramme der FeCh Familie beinhalten das Auftreten von verschiedenen Phasen wie Magnetismus oder Supraleitung. Während die Muttersubstanzen im Allgemeinen antiferromagnetisch geordnet sind, werden die Materialien in der Regel supraleitend nachdem sie eine Region der Koexistenz zwischen Supraleitung und Magnetismus durchlaufen haben. In dieser Arbeit wurde das FeCh System nur durch Änderungen am Kristallgitter manipuliert. Hierfür wurde entweder hydrostatischer Druck in einer Druckzelle oder chemischer Druck durch Substitution von Se durch das isovalente Te angelegt ohne zusätzliche Ladungsträger in das System einzufügen. Die Myon Spin Rotation/Relaxation/ Resonanz (μSR) Technik in Kombination mit ac und dc Magnetisierungsmessungen ist eine optimale Kombination um den supraleitenden und magnetischen Zustand zu untersuchen und wie die Beiden miteinander interagieren im Sinne von Koexistieren und/oder in Konkurenz zueinander stehen. Dabei stellte sich heraus, dass das System extrem sensitiv auf Druck reagiert. FeSe$_{1-x}$ ist bei Umgebungsdruck ausschliesslich supraleitend und nicht magnetisch. Indem Druck angelegt wird, erhöht sich die supraleitende Sprungtemperatur und das System zeigt einen der grössten Druckeffekte, die in der Natur bekannt sind. Überraschenderweise wird es zusätzlich bei hohen Drücken magnetisch und Magnetismus und Supraleitung koexistieren in FeSe$_{1-x}$ auf atomaren Längenskalen. Einen ähnlichen Effekt kann man durch die Substitution von Se durch das isovalente Te beobachten. Zunächst steigt die supraleitende Sprungtemperatur an, anschliessend wird das System supraleitend und magnetisch zugleich und schliesslich tritt es in eine Phase, die ausschliesslich antiferromagnetisch ist. Diese Beobachtungen lassen den Schluss zu, dass in der FeCh Familie das Kristallgitter eine aussergewöhnlich wichtige Rolle zu spielen scheint, da bereits kleinste Änderungen am System zu drastischen Effekten an den elektronischen Eigenschaften führen.

Um Hinweise auf den mikroskopischen Paarungsmechanismus, welcher zur Supraleitung in den eisenbasierten Hochtemperatursupraleitern führt, zu finden wurde die Temperaturabhängigkeit der suprafluiden Dichte und die daraus resultierende Struktur der Energielücke untersucht. Es stellt sich herraus, dass alle eisenbasierten Supraleiter einschliesslich FeCh mehrere Energielücken aufweisen. Interessanterweise ist das Verhältnis der Energielücke zur Sprungtemperatur in allen eisenbasierten Supraleitern ähnlich zueinander. Ein weiterer Hinweis auf den Paarungsmechanismus liefern Isotopenaustausch Experimente. Sie zeigten, dass die supraleitende Sprungtemperatur von der Isotopenmasse abhängt, was eindeutig die Wichtigkeit des Gitters zum Paarungsmechanismus hervorhebt.

Contents

Abstract	v
Zusammenfassung	ix
1 Introduction	**1**
2 Basic properties of superconductors	**5**
2.1 Introduction to superconductivity	5
2.2 Energy gap and excitation spectrum	7
3 Muon spin rotation/relaxation/resonance (μSR)	**9**
3.1 Introduction	9
3.2 Principle of μSR	11
3.3 Muons in materials	13
3.3.1 Muons in magnetic materials	13
3.3.2 Muons in superconducting materials	15
4 The FeCh system	**19**
4.1 Fe-based superconductors	19
4.2 Synthesis of FeSe$_{1-x}$	23
4.3 Hydrostatic pressure effect	25
4.4 Chemical pressure effect and role of Fe	35
4.5 Related publications to Chapter 4	43
4.5.1 Paper I: Synthesis, crystal structure, and chemical stability of the superconductor FeSe$_{1-x}$	43
4.5.2 Paper II: Pressure Induced Static Magnetic Order in Superconducting FeSe$_{1-x}$	44

	4.5.3	Paper III: Evolution of Two-Gap Behavior of the Superconductor FeSe$_{1-x}$ 45
	4.5.4	Paper IV: Coexistence of incommensurate magnetism and superconductivity in Fe$_{1+y}$Se$_x$Te$_{1-x}$. 46
	4.5.5	Paper V: Tuning the superconducting and magnetic properties in Fe$_y$Se$_{0.25}$Te$_{0.75}$ by varying the Fe-content 47
	4.5.6	Paper VI: Anisotropic superconducting properties of single-crystalline FeSe$_{0.5}$Te$_{0.5}$ 48

5 Isotope effect   **51**
    5.1   Isotope effect in the Fe-based superconductors 53
    5.2   Related publications to Chapter 5 58
         5.2.1   Paper I: Iron isotope effect on the superconducting transition temperature and the crystal structure of FeSe$_{1-x}$ 58
         5.2.2   Paper II: Intrinsic and structural isotope effects in Fe-based superconductors 59

6 Conclusion and Outlook   **61**

Bibliography   **65**

1 Introduction

After the liquefaction of helium by H. K. Onnes in 1908 [1] he investigated the resistivity of metals at low temperature. In doing so, he discovered in 1911 that the electrical resistivity of mercury drops to zero below a critical temperature $T_c \simeq 4.19\,\mathrm{K}$ [2]. First he called the phenomenon "supraconductivity" and only later adopted the term "superconductivity". In the following years more superconducting materials as e.g. lead with a $T_c \simeq 7\,\mathrm{K}$ or niobium nitride with $T_c \simeq 16\,\mathrm{K}$ were found.

An important step in understanding superconductivity was made in 1933, when W. Meissner and R. Ochsenfeld discovered an effect called the "Meissner effect" [3]. In a weak magnetic field it is expelled completely from the inside of the superconductor below T_c, independent on the cooling history. Accordingly, superconductors are not only perfect conductors but also perfect diamagnets which lead to the conclusion that superconductivity is a thermodynamic phase. Shortly after, in 1935 F. and H. London developed the first phenomenological theory describing the thermodynamic properties of a superconductor [4, 5]. In this theory the zero resistance and the Meissner effect were described. Fifteen years later, in 1950 V. Ginzburg and L. Landau developed the Ginzburg-Landau theory that combined Landau's theory of second order phase transitions with a wave function that characterizes the superconducting state [6]. As was shown later by A. Abrikosov [7] it was the first theory specifying type I and type II superconductors by introducing the coherence length. It had great success explaining the macroscopic properties of a superconductor.

A major role to describe the mechanism behind superconductivity played the finding of the isotope effect in 1950 when E. Maxwell [8] and C. A. Reynolds *et al.* [9] discovered it in mercury at about the same time independently. It indicates that the electron-phonon interaction is a main ingredient of the mechanism leading to superconductivity. Actually, this was afterwards described by the first widely-accepted theory explaining the effect of superconductivity that was advanced in 1957 by J. Bardeen, L. Cooper, and J. Schrieffer [10, 11]. The so-called BCS theory explains superconductivity close to absolute zero, whereas the supercurrent consists of Cooper pairs, *i.e.* pairs of electrons that are interacting with each other through the exchange of phonons. It was awarded the Nobel prize in 1972. In 1959 shortly after the development of the BCS-theory L. Gor'kov showed that the microscopic BCS theory reduces to the phenomenological Ginzburg-Landau theory close to T_c [12]. After the theoretical description of superconductivity by the BCS theory and the knowledge, what properties a superconductor should have to reach a high T_c (high density of states at the Fermi level, a strong electron phonon interaction, and a high Debye temperature) T_c increased only marginally to the record of 23 K found in Nb_3Ge in 1973. However, the BCS theory reveals that the maximum value of T_c can not be higher than 30 K [13].

In 1986 the striking discovery of the cuprate high-temperature superconductivity in Ba-La-Cu-O by J. G. Bednorz and K. A. Müller attracted the attention of the scientific community [14]. This lead to the class of the cuprate high temperature (HTC) superconductors which reached superconducting transition temperatures higher than 77 K, the boiling point of nitrogen. Till now the highest T_c is 135 K at ambient pressure [15] and reaches 164 K at high pressure [16]. The name HTC superconductors came up because T_c is higher than the supposed BCS value of 30 K and the fraction T_c/T_F (T_F Fermi temperature) is larger than in elemental, or classical superconductors as mercury or lead. However, after 25 years of research the origin of high temperature superconductivity is still not clear, and the HTC superconductors cannot be explained by the BCS theory. But there are hints that instead of electron-phonon interaction as in the conventional superconductors it is rather a polaronic mechanism [17, 18, 19] and instead of pure *s*-wave, rather mixed *d*- plus *s*-wave pairing is substantial [19, 20].

Over a decate later, in 2001 the binary MgB$_2$ was found to be superconducting at $T_c \simeq 39$ K [21]. It still is a HTC superconductor, but compared to the cuprates it is a more conventional one in the sense of electron-phonon mediated high temperature superconductivity in which not only one superconducting gap opens. It was shown that MgB$_2$ is a two-gap superconductor [22]. Again the field of superconductivity gained major interest when in 2008 the iron-based HTC superconductors were discovered with T_c reaching values up to 56 K [23]. Also here the microscopic origin of superconductivity is not explained yet. But it seems that the iron-based superconductors are again multi-gap superconductors with a complicated gap structure, whereas the exact symmetry of the gap still is under discussion [24].

For understanding high temperature superconductivity it is important to compare the different classes with each other. All of the high temperature superconductors share some important features like e.g. the layered structure and the resulting pronounced, temperature dependent anisotropic behavior, a small coherence length, competing order parameters, and possible multi-gap superconductivity. But there are also important differences between the classes as the iron-based HTC superconductors have metallic parent compounds while the parent compounds of the cuprates are in general insulators. Furthermore, the anisotropy of the iron-based materials is in general lower than the one of the cuprates and the temperature dependences of the anisotropy parameters of the magnetic penetration depth and the coherence length are opposite in the iron-based superconductors compared to MgB$_2$. Another important difference is the order parameter that is $d+s$-wave in the cuprates, $s+s$ in MgB$_2$ and most probably $s_\pm$ in the iron-based superconductors, whereas in the latter Fermi surface nesting seems to play a major role. Hence, to understand the mechanism leading to superconductivity and the fundamental question whether the origin is similar, it is important to investigate the superconducting and magnetic properties and the interplay of both of them of these systems in detail. Especially in the iron-based HTC superconductors there are large regions of coexistence of superconductivity and magnetic order in the phase diagrams. Among

Chapter 1. Introduction

the iron-based superconductors the FeCh (Ch = chalcogenide) system is an ideal system to study, since it is the simplest one due to the layered crystallographic structure with no separating layers between the superconducting ones (see Chapter 4) and the similarity of the Fermi-surface topology with that of other iron-based superconductors [25]. Muon spin rotation (μSR) in combination with magnetization measurements is an optimal tool to investigate the superconducting properties of the system and the interplay of superconductivity and magnetism.

In Chapter 2 a general introduction to some important phenomenological parameters and phenomena of superconductivity is given, forming a basis for the subsequent discussion of the techniques and results. Chapter 3 gives an overview on the μSR technique and its sensitivity to characteristic parameters for the different classes of materials. The following chapter 4 presents the superconducting and magnetic properties of the FeCh system and the preparation of the samples. Furthermore, the influence of hydrostatic and chemical pressure as well as the influence of excess Fe on the system is discussed. Chapter 5 deals with the iron isotope effect on the superconducting transition temperature in FeSe$_{1-x}$. As already mentioned above, the isotope effect played a major role in finding a theory for conventional superconductors. Also in the iron-based superconductors it might give a major hint at the pairing mechanism.

2 Basic properties of superconductors

Superconductors exhibit two main characteristics below the critical temperature (superconducting transition temperature) T_c: They are perfect conductors, meaning that the resistivity is unmeasurable small and they are perfect diamagnets, that expel a magnetic field up to a critical field H_c completely from their interior which is known as the Meissner effect. In this chapter the basics of the theory of superconductors are introduced. The interested reader is referred to more comprehensive textbooks [26, 27, 28].

2.1 Introduction to superconductivity

A mathematical model to describe superconductivity is based on Landau's theory of second order phase transitions [29]. Motivated by the London theory, showing that superconductors behave as though governed by a macroscopic wave function, Ginzburg and Landau introduced a complex, spatially varying order parameter. It is characterized by a complex pseudo-wave function $\psi(r)$ with $|\psi(r)|^2 = n_s(r)$, whereas n_s is the superconducting carrier density. According to the Landau theory the total free energy f of a system can be reduced by entering a new phase. Following the Ginzburg-Landau postulate the free energy density f can be expressed in the superconducting state of a spatially inhomogeneous superconductor in an applied magnetic field in a series of the form [6]:

$$f(\mathbf{H}) = f_{n0} + \alpha |\psi|^2 + \frac{\beta}{2} |\psi|^4 + \frac{1}{2m^*} \left| \left(\frac{\hbar}{i} \nabla - e^* \mathbf{A} \right) \psi \right|^2 + \frac{\mu_0 \mathbf{H}^2}{2}. \quad (2.1)$$

Chapter 2. Basic properties of superconductors

Here m^* denotes the effective mass and e^* the charge of the superfluid carriers, i.e. $m^* = 2m$ and $e^* = 2e$ for Cooper pairs in metals, f_{n0} is the normal state free energy, $\psi = |\psi| \exp[i\varphi]$ the complex, macroscopic order parameter of the superconducting phase, and α and β the expansion coefficients from the Landau theory of phase transitions.

Integration of the free energy over the whole sample volume and minimization by variation of ψ and $\mathbf{A}$ leads to the Ginzburg-Landau equations:

$$\alpha\psi + \beta |\psi|^2 + \frac{1}{2m^*}\left(\frac{\hbar}{i}\nabla - e^*\mathbf{A}\right)^2 \psi = 0 \quad (2.2)$$

$$\mathbf{J}_s = \frac{e^*}{m^*}|\psi|^2 \left(\hbar\nabla\varphi - e^*\mathbf{A}\right) = e^* |\psi|^2 \mathbf{v}_s \quad (2.3)$$

Here $\mathbf{v}_s$ denotes the average carrier velocity and $\mathbf{J}_s$ the density of circulating supercurrents. The Ginzburg-Landau theory contains two characteristic length scales. The 2nd Ginzburg-Landau equation [Eq. (2.3)] represents the expression for the supercurrent density and is analogous to

$$\mathbf{J}_s = -\frac{n_s^* e^{*2}}{m_e^*}\mathbf{A} \quad (2.4)$$

obtained within the generalized London theory [4]. The length scale derived from that equation is the London penetration depth

$$\lambda_\mathrm{L} = \sqrt{\frac{m_e^*}{\mu_0 n_s^* e^{*2}}} \quad (2.5)$$

which describes perfect conductivity, the Meissner effect, and flux quantization.

The 1st Ginzburg-Landau equation [Eq. (2.2)] describes implicitly the spatial variation of $\psi(r)$, yielding an equation that has the dimension of length:

$$\xi_\mathrm{GL} = \sqrt{\frac{\hbar^2}{2m^*\alpha}} \quad (2.6)$$

This is the second characteristic length scale in the Ginzburg-Landau theory, called the Ginzburg-Landau coherence length ξ_GL.

The ratio of the two characteristic length scales is denoted as the Ginzburg-Landau parameter

$$\kappa = \frac{\lambda_\text{L}}{\xi_\text{GL}} \quad (2.7)$$

that is independent of temperature in classical superconductors. It allows to distinguish between two types of superconductors: type-I and type-II superconductors [26]. In the introduction to this chapter it was mentioned that superconductors behave as perfect diamagnets expelling the applied field H_appl completely up to a critical field H_c where superconductivity is suppressed. This behavior is called the Meissner effect but it is valid only for type-I superconductors for which $\kappa < 1/\sqrt{2}$. However, there are also type-II superconductors for which $\kappa > 1/\sqrt{2}$ and that show a completely different behavior. As long as H_appl is smaller than the lower critical field H_c1, type-I and type-II superconductors behave the same. But as soon as $H_\text{appl} > H_\text{c1}$ the magnetic field partly penetrates a type-II superconductor and forms the so-called mixed state or Shubnikov phase. Only when H_appl reaches the value of the upper critical field H_c2 that is usually much higher than H_c in a type-I superconductor superconductivity is suppressed. Generally, in the mixed state a regular arrangement of normal conduction regions containing a single flux quantum $\Phi_0 = \pi\hbar/e$ is formed, denoted as the flux line lattice.

2.2 Energy gap and excitation spectrum

Bardeen, Cooper, and Schrieffer [10, 11] proposed in their BCS theory a weak attractive force between electrons, that is caused by the electron-phonon interaction. This causes an instability at the Fermi ground state of the electron gas and bound pairs of electrons are formed that have opposite momentum and opposite spin. Such an electron pair is called a Cooper-pair which has the spacial size of the order of the coherence length ξ_0 [30]. They are much larger than the interparticle distance and thus they are highly overlapping. The formation of the Cooper pairs leads to a gain of energy which gives rise to a gap at the Fermi energy. The following equation determines the temperature dependence of the

Chapter 2. Basic properties of superconductors

superconducting gap parameter. It follows from the BCS theory and has to be solved numerically for $T > 0$ [26]:

$$\frac{1}{N(0)V} = \int_{-\hbar\omega_D}^{\hbar\omega_D} \frac{\tanh\frac{\sqrt{\xi_k^2+\Delta_k^2}}{2k_BT}}{\sqrt{\xi_k^2+\Delta_k^2}} \mathrm{d}\xi_k. \qquad (2.8)$$

Here, $(\xi_k^2 + \Delta_k^2)^{1/2}$ is the excitation energy of a fermion quasi particle, whereas $\xi_k = E - E_F$ (E_F is the Fermi energy) and Δ_k is the energy gap, $N(0)$ and V are the density of states at the Fermi level and the electron-phonon interaction strength, respectively.

For $T = 0$ it can be evaluated that the gap is comparable to $k_B T_c$:

$$\frac{2\Delta(0)}{k_B T_c} = 3.5. \qquad (2.9)$$

However, experimentally it is found in classical superconductors that the values of $2\Delta(0)$ range from $3k_B T_c$ to $4.5 k_B T_c$.

For weak coupling superconductors $\Delta(T)/\Delta(0)$ is a monotonically decreasing function of T/T_c from 1 to 0. One sees from Eq. (2.8) that near $T = 0$ the temperature dependence of $\Delta(T)$ is very weak. This implies that the gap is nearly constant close to $T = 0$ and changes only when a significant number of quasi-particles are thermally excited. At the same time at $T \approx T_c$ the superconducting gap closes relatively fast by following the relation:

$$\Delta(T) \simeq 1.74\Delta(0)\sqrt{1 - \frac{T}{T_c}} \qquad (2.10)$$

One finds that the temperature dependence of the density of charge carriers n_s depends on the temperature evolution of the gap parameter Δ^2. The calculations yield that on the one hand $\Delta(0)$ is half of the energy gap of the superconductor and on the other hand it also determines the number of Cooper pairs [26].

3 Muon spin rotation/relaxation/resonance (μSR)

The acronym μSR stands for "muon spin rotation/relaxation/resonance". This technique, which has found a wide application in solid states physics, makes possible to study magnetism, superconductivity, diffusion processes, kinetics and molecular dynamics, and semiconductivity. In this chapter a brief introduction to the μSR technique is given, and its application to the study of magnetic materials and the internal field distributions in type-II superconductors is discussed. The interested reader is referred to more comprehensive textbooks [31, 32].

3.1 Introduction

The muon is a particle belonging to the family of the leptons with an average lifetime of $\tau_\mu \simeq 2.2$ μs. The muon mass is about $1/9^\text{th}$ of the proton mass or, alternatively, about 200 times the electron mass. The muon is the decay product of pions that are produced upon bombarding nucleons with other nucleons. At the Paul Scherrer Institute (PSI Villigen, Switzerland) this is performed by accelerating protons with a 590 MeV cyclotron that hit a carbon target. There, the following reactions are taking place:

$$\begin{aligned} \text{p}+\text{p} &\to \text{p}+\text{n}+\pi^+ & \text{p}+\text{n} &\to \text{p}+\text{n}+\pi^0 \\ &\to \text{d}+\pi^+ & &\to \text{p}+\text{p}+\pi^- \\ &\to \text{p}+\text{p}+\pi^0 & &\to \text{n}+\text{n}+\pi^+ \end{aligned}$$

All the resulting pions (π^+, π^-, and π^0) have a spin $S = 0$. Solely the π^+ and π^- are decaying into muons whereas the π^0 decays into two γ

rays:

$$\begin{aligned} \pi^+ &\to \mu^+ + \nu_\mu \\ \pi^- &\to \mu^- + \bar{\nu}_\mu \\ \pi^0 &\to \gamma + \gamma \end{aligned}$$

Due to the spin conservation, and since the pions have a spin 0 and the neutrinos a spin 1/2, the spin of the muons is equal to 1/2. Furthermore, as solely left-handed ν_μ exist, the pion decay violates parity leading to the production of 100 % spin polarized muons. Due to the negative helicity of the neutrino, the muon spin points to opposite direction of its momentum.

For µSR experiments in condensed matter physics, positively charged μ^+ are used. They come to rest at interstitial lattice positions as a free particle after the thermalization process. On the other hand, negative charged muons μ^- form an excited muonic atom where the muon behaves as a heavy electron. After an averaged lifetime of $\tau_\mu \simeq 2.2\,\mu s$, the muon μ^+ decays into a positron e$^+$, an electron-neutrino ν_e, and a muon-antineutrino $\bar{\nu}_\mu$, as the lepton number and the charge must be conserved:

$$\mu^+ \to e^+ + \nu_e + \bar{\nu}_\mu$$

This decay, as the pion decay, occurs via the weak interaction, and as the neutrinos have a negative helicity and the antineutrinos a positive, the decay positron tends to be emitted along the muon spin direction at the time of the decay.

Roughly speaking, there are three key properties of the muon making µSR possible: (i) the muon is 100% spin polarized, (ii) the positron is preferentially emitted along the direction of the muon spin at decay time, and (iii) the muon has a magnetic moment and its spin precesses around a magnetic field with the Larmor frequency.

3.2 Principle of µSR

The µSR method is based on the observation of the time evolution of the muon spin polarization $P(t)$ of muons, that are implanted into a sample. The basic principle of a µSR experiment is illustrated in Fig. 3.1. At the time of the muon implantation into the sample, a clock triggered by the muon detector is started. In the local magnetic field the muons spin starts to precess with the Larmor frequency ω_L until it decays and emits a positron. The latter is then detected by one of the positron-detectors which stops the clock. As a result a histogram as a function of time is generated for the forward ($N_F(t)$, forward with respect to the spin) and the backward ($N_B(t)$) detectors:

$$N_{F(B)}(t) = N_0 \exp[-t/\tau_\mu] \cdot (1 + A_{max} P(t) \cdot \hat{n}_{F(B)}), \qquad (3.1)$$

where

$$P(t) = \frac{\langle I(t) \cdot \hat{n} \rangle}{|I(0)|} \qquad (3.2)$$

is the polarization function with the unit vector $\hat{n}$ with respect to the direction of the incoming muon spin polarization $I(0)$. To get the time evolution of the muon polarization either the exponential decay component due to the muon decay can be fitted or the asymmetry can be obtained from:

$$A(t) = A_{max} P(t) = \frac{N_F(t) - \beta N_B(t)}{N_F(t) + \beta N_B(t)}. \qquad (3.3)$$

Here, the function $A(t)$ is the asymmetry, that contains the information about the physics whereas A_{max} depends on different experimental factors, such as the detector solid angle, efficiency, absorption, and scattering of positrons in the material. The values typically lie between 0.25 and 0.3. The parameter β is introduced to take into account the different efficiency of the positron-detectors and needs to be determined by calibration.

Two different types of muon beams are used for µSR: continuous beams like at PSI and TRIUMF (Canada) and pulsed beams like the ones at

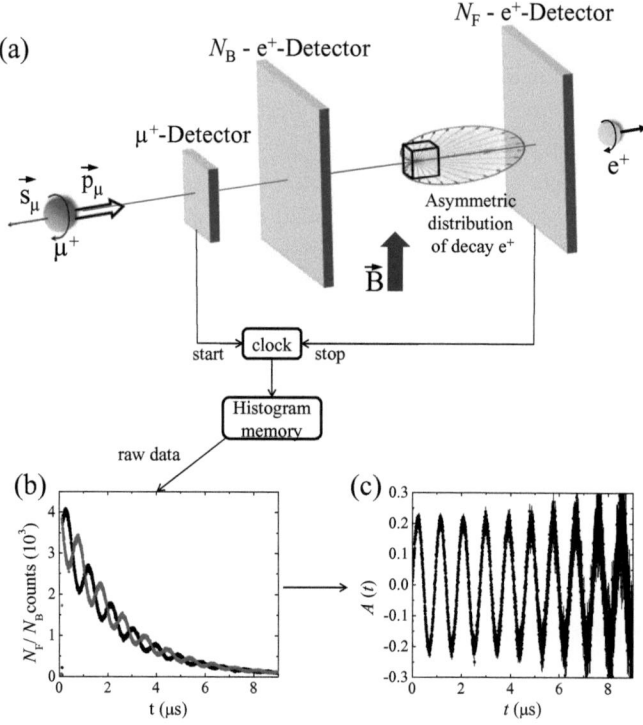

Figure 3.1: Principle of a μSR experiment as explained in the text. (a) shows the simplified experimental setup of a beamline: A spin polarized muon is implanted in the sample that is placed between the forward and the backward positron detector. In the transverse field (TF) configuration the external field B_{ext} is applied perpendicular to the initial muon spin polarization, whereas in longitudinal field μSR experiments the external field is applied parallel to the muon spin polarization. In zero-field μSR no field is applied, and the muon precesses only in the local magnetic field of the sample B_{loc}. A clock is started at the time the muon enters the muon detector/sample and stopped again as soon as the decay positron is detected. (b) The number of detected positrons as a function of time for both individual detectors. The histograms still contain the natural life-time decay of the muons. (c) The asymmetry signal obtained from the forward and backward positron detector with the help of Eq. 3.3.

ISIS (U.K.) and J-PARC (Japan). In a continuous beam a nearly continuous source of spin-polarized muons are implanted one at a time into the sample and for each individual muon its decay positron is counted. That limits the muon implantation rate since the corresponding muon/positron events need to be clearly differentiate from another pair. However, the paramount advantage of this type of muon beam is the small time resolution of ≈ 100 ps which give the possibility to investigate large magnetic fields and fast relaxing signals. On the other hand, in a pulsed beam a bunch of muons are implanted into the sample at the same time, with the conditions that the muon pulse must be considerably shorter than the lifetime of the muons and that the pulse repetition period must be much longer than the muon lifetime. A pulsed beam cannot compete with a continuous beam measuring fast relaxing μSR signals or μSR spectra in high magnetic fields, since the pulses have typically a width of 80 ns. The advantages on the other hand are, that one can use the entire amount of incoming muons, a lower background due to the absence of accidental double events, and by synchronizing the muon beam with e.g. an RF field one has the possibility to study resonance effects.

3.3 Muons in materials

Two different magnetic field configurations for μSR experiments are used: (i) the transverse-field (TF) μSR and (ii) the longitudinal (LF) and zero-field (ZF) μSR. In the TF configuration an external magnetic field $\vec{B}_{\text{ext}}$ is applied perpendicular to the original muon polarization, whereas in LF μSR the field is applied along the initial muon spin direction, and in ZF μSR no field is applied.

3.3.1 Muons in magnetic materials

Muons are ideal local probes to study problems in magnetism like determining the internal field distributions and the magnetic ground state of systems. This is performed by ZF μSR (zero applied field, see above).

Chapter 3. Muon spin rotation/relaxation/resonance (μSR)

For the muons stopping in magnetically ordered systems, their spins precess in the local field B_{loc} with the Larmor frequency $\omega_\mu = \gamma_\mu \cdot B_{\text{loc}}$ ($\gamma_\mu = 2\pi \cdot 135.5\,\text{MHz/T}$ is the gyromagnetic ratio of the muon), yielding a precessing signal directly proportional to B_{loc}. Due to their large magnetic moment the muons are very sensitive to small magnetic fields ($\simeq 10^{-5}$ T, i.e. values of the order of magnetic fields created by nuclear moments). As the muons are stopping at well defined crystallographic sites, but randomly in the sample, the μSR technique can be utilized to check the coexistence of different types of ground states at the microscopic level. If different ground states are present in a sample, they will be characterized in the μSR spectrum by different components with amplitudes proportional to the corresponding volume fractions. This makes the technique extremely useful in the case where the sample consists of multiple phases or if an incomplete magnetic ordering occurs.

To get a flavor of why the muon is able to study randomness and dynamics in magnetic materials some aspects of spin precession are hereafter considered. If the magnetic field at the muon site is at an angle ϑ to the initial muon spin direction, the muon spin will precess around the direction of the field along a cone with angular aperture ϑ. The decay positron asymmetry is then given by:

$$A(t) = A_{\max} \left[\cos^2 \vartheta + \sin^2 \vartheta \cos\left(\gamma_\mu B_{\text{loc}} t\right) \right]. \qquad (3.4)$$

In the case of a polycrystal, the direction of the local internal field is random with respect to the initial polarization. Averaging over all directions yields

$$A(t) = A_{\max} \left[\frac{1}{3} + \frac{2}{3} \cos\left(\gamma_\mu B_{\text{loc}} t\right) \right]. \qquad (3.5)$$

If the internal field is not constant, but is Gaussian distributed around zero with a width Δ/γ_μ, one obtains the well-known formula

$$A(t) = A_{\max} \left[\frac{1}{3} + \frac{2}{3} \left(1 - \Delta^2 t^2\right) \exp\left[-\Delta^2 t^2/2\right] \right], \qquad (3.6)$$

developed by Kubo and Toyabe [33]. This relaxation function falls from its initial value to a minimum and recovers to an average value $A(t \to$

$\infty) = 1/3A_{\text{max}}$. Obviously, the form of the internal field distribution will strongly affect the form of the observed μSR time spectrum, e.g. if the magnetic order is incommensurate with the crystal lattice, the muon spin relaxation follows a Bessel function [34, 35].

If there is a variation of the field strength, different muons will precess with different frequencies, leading to a dephasing so that the oscillations will be damped. The bigger this variation is, the larger is the damping until the oscillations vanish. However, such an effect could also arise from fluctuations either of the internal field or due to muon diffusion. The origin of the vanishing of the oscillations can be determined by so called longitudinal field (LF) measurements, where the field is applied parallel to the initial muon spin direction [36]. Already relatively low magnetic fields ($B_{\text{ext}} \leq 10 \times B_{\text{loc}}$) have a large effect in the case of static magnetism or weak dynamics, but less effect if the dynamics are fast.

3.3.2 Muons in superconducting materials

Using the μSR technique important length scales of superconductors can be measured, namely the magnetic penetration depth λ and the coherence length ξ (see Chapter 2) [37]. If a type II superconductor is cooled below T_c in an applied magnetic field $H_{c1} < H < H_{c2}$ a vortex lattice is formed which in general is incommensurate with the crystal lattice and the vortex cores will be separated by much larger dimensions than those of the unit cell. Because the implanted muons stop at given crystallographic sites, they will randomly probe the field distribution of the vortex lattice.

Such measurements need to be performed in an applied field perpendicular to the muon spin (TF configuration). In the normal state, all muons precess with the frequency $\omega = \gamma_\mu B_{\text{ext}}$ and the field distribution is ideally a δ-peak (some broadening may occur due to the nuclear moments leading to a depolarization rate σ_{nm}, see below). In the superconducting state, however, the muons will sample the field distribution created by the vortex lattice. Muons stopping close to a vortex core experience larger fields than the ones stopping in between vortices. The field distribution sensed by the muons will result in a damping of the

Chapter 3. Muon spin rotation/relaxation/resonance (μSR)

muon precession signal. The larger the penetration depth of the superconductor (characterizing the decay of the field around a vortex), the smaller is the magnetic field variation and thus the damping. The field distribution created by the vortices is highly asymmetric as shown in Fig. 3.2. It turns out, that a sum of oscillating signals with Gaussian damping is a reasonable approximation if the probed field distribution is due to an ordered or only weakly disordered vortex lattice [38]. In this case the muon asymmetry is described by:

$$A(t) = \sum_{i=1}^{n} A_i \exp[-\sigma_i^2 t^2/2] \cos(\gamma_\mu B_i t + \varphi). \tag{3.7}$$

Here A_i are the asymmetries, σ_i the Gaussian depolarization rates, and B_i the average fields of the components i. φ is the common initial phase of the muon spins with respect to the muon detector. The total asymmetry $A = \sum_{i=1}^{n} A_i$ and the second central moment of the corresponding field distribution is given by:

$$\langle \Delta B^2 \rangle = \frac{\sigma^2}{\gamma_\mu^2} = \frac{1}{A} \sum_{i=1}^{n} A_i \left(\sigma_i^2/\gamma_\mu^2 + (B_i - \langle B \rangle)^2 \right), \tag{3.8}$$

with the corresponding first moment of the field distribution:

$$\langle B \rangle = \frac{1}{A} \sum_{i=1}^{n} A_i B_i. \tag{3.9}$$

The depolarization of the muon spin ensemble is assumed to be caused only by the inhomogeneous field distribution of the vortex lattice and random nuclear moments. The contribution from the vortex state is given by $\sigma_{sc}^2 = \sigma^2 - \sigma_{nm}^2$, where σ_{nm} is the depolarization of the nuclear moments obtained above T_c. For further interpretation, the magnetic penetration depth λ is evaluated from the depolarization rate σ_{sc} by the relation [39]:

$$\sigma_{sc} = 4.83 \times 10^4 \, (1-b) \left[1 + 3.9 \, (1-b)^2 \right]^{\frac{1}{2}} \lambda^{-2}, \tag{3.10}$$

where $b = \langle B \rangle / B_{c2}$ is the reduced upper critical field of the type-II superconductor. Thus, the relaxation rate of the observed precession

3.3. Muons in materials

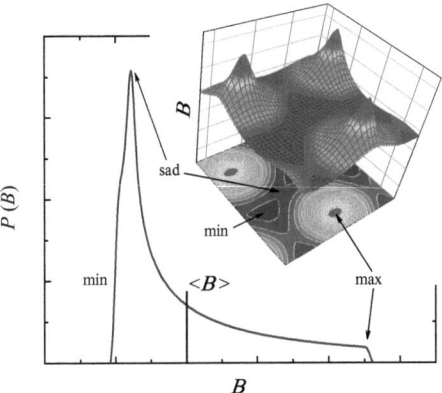

Figure 3.2: Field distribution $P(B)$ of a perfect regular vortex lattice. The inset corresponds to the spatial distribution.

signal can be used to determine the magnetic penetration depth. The determination of λ by μSR can be considered as a bulk technique, in contrast to many techniques that probe λ only from the surface.

A different, more advanced approach is to describe the spatial field distribution of the magnetic field in the vortex state by a Fourier series:[40, 41, 38]

$$B(\mathbf{r}) = \langle B \rangle \sum_{\mathbf{G}} \exp\left[-i\mathbf{G}\mathbf{r}\right] B_{\mathbf{G}}(\lambda, \xi), \qquad (3.11)$$

where $\mathbf{r}$ is the vector coordinate in the plane perpendicular to the applied field, $\mathbf{G}$ the reciprocal lattice vectors of a two-dimensional hexagonal vortex lattice, and $B_{\mathbf{G}}$ the Fourier components. The field distribution $P(B)$ is probed by the muons by random sampling of $B(\mathbf{r})$ leading to the following asymmetry function:

$$A(t) = A_{\max} \exp\left[-\sigma_{\mathrm{g}}^2 t^2/2\right] \int P(B) \cos\left(\gamma_\mu B t + \varphi\right) \mathrm{d}B. \qquad (3.12)$$

Here the Gaussian prefactor takes the broadening of the field distribution by e.g. nuclear dipole fields or weak pinning into account [42], whereas in

Chapter 3. Muon spin rotation/relaxation/resonance (μSR)

the above described Gaussian approximation only the nuclear moments are taken into account and one assumes a perfect vortex lattice.

4 The Fe*Ch* system

This chapter is focused on the superconducting and magnetic phases, their properties, and their competition, coexistence, and interplay in the Fe*Ch* system (*Ch* = chalcogen S, Se, Te), which is the most simple among the Fe-based superconductors discovered in 2008 [23]. First an introduction to the Fe-based superconductors is given, followed by a detailed description of the preparation procedure. Then the hydrostatic as well as the chemical pressure effect on $FeSe_{1-x}$ and finally the influence of excess Fe on the physical properties is discussed.

4.1 Fe-based superconductors

In the year 2008 Kamihara *et al.* discovered superconductivity in the iron based compound $LaFeAsO_{1-x}F_x$ with a superconducting transition temperature $T_c \simeq 26\,K$ [23]. It is worth to mention, that superconductivity in the isostructural $LaFePO_{1-x}F_x$ was observed earlier by the same group in 2006 with $T_c \simeq 5\,K$ [43]. But at that time even the fact that Fe-based or Fe-containing materials become superconducting was not noteworthy, as e.g. the Fe-containing U_6Fe [44] or the skutterudites XFe_4P_{12} [45, 46] with T_c's up to $7\,K$ were already known. Yet only the later discovery captured the imagination of physicists and chemists worldwide as high transition temperatures can be reached although elemental iron is a ferromagnet. By applying pressure in the $LaFeAsO_{1-x}F_x$ compound T_c increases even further to $\sim 43\,K$ [47]. The compound grows in the primitive tetragonal ZrCuSiAs-type (1111-type) structure [48] and contains FeAs layers with Fe atoms in a square planar lattice arrangement tetrahedrally coordinated by As atoms. The FeAs layers alternate with the LaO layers along the *c*-axis.

Chapter 4. The FeCh system

Remarkably, by replacing the non-magnetic La by magnetic rare earth metals such as Sm, Nd, or Ce, the transition temperature T_c even increased further to its current highest value in the Fe-based compounds of $\simeq 56\,\text{K}$ in SmFeAsO$_{1-x}$F$_x$ [49, 50, 51, 52, 53]. The materials with the highest T_c's differ from the ones with a lower T_c in a sense that application of pressure decreases T_c in the materials with the highest T_c. This leads to the conclusion that there is an optimal ion height above the Fe-plane [54].

Shortly after the discovery of the high T_c's in the 1111 family of Fe-based superconductors, other superconducting families with similar FeAs layers were found. The Ba$_{1-x}$K$_x$Fe$_2$As$_2$ compound in the body-centered-tetragonal ThCr$_2$Si$_2$ (122-type) structure shows so far a maximum of $T_c \simeq 38\,\text{K}$ for $x \simeq 0.4$ [55]. Subsequently, by replacing the Ba/K by other elements (e.g. Cs, Sr, Na, Rb, Eu) various related compounds were discovered [53, 56, 57, 58, 59]. They are, however, separated only by a single layer consisting of one atom. Other similar superconducting materials such as LiFeAs (111-type) containing FeAs layers were also reported during that time [60, 61]. Meanwhile even more complicated systems as the (Fe$_2$As$_2$)(Ae_4M_2O$_6$) (22426, Ae = alkaline earth metal and M = transition metal) and (Fe$_2$As$_2$)(Ae_3M_2O$_5$) (22325) were detected [62, 63].

Even the binary compound α-FeSe$_{1-x}$ (11-type) with a layered structure was found to be superconducting at $T_c \simeq 8\,\text{K}$ [64]. From the first point of view it could be argued that FeSe$_{1-x}$ is not a high temperature superconductor as T_c is that low but after a short time it was found upon studying the pressure effect, that T_c is very sensitive on chemical (substitution of Se by isovalent Te or S) or hydrostatic pressure and reaches values of 36 K at $p \simeq 9\,\text{GPa}$ [65, 66, 67, 68].

Very recently superconductivity at about 30 K was reported in the FeSe-layer compound K$_{0.8}$Fe$_{2-x}$Se$_2$ [69] by intercalating potassium in a solid state reaction between the FeSe layers whereas the substitution of K by e.g. Cs or Rb did not change T_c [69, 70, 71]. This system is isostructural to the 122-type materials and thus might allow a direct comparison of the Se-based 122- to the As-based 122-type of Fe-based superconductors.

4.1. Fe-based superconductors

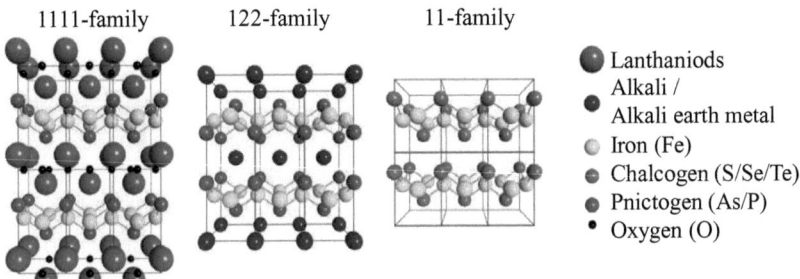

Figure 4.1: Comparison of the crystal structures of the 1111-type, the 122-type, and the 11-type. All crystal structures consist of the superconducting M_2X_2 layers separated by insulating layers. Each of the M_2X_2 layers consists of a square lattice of M atoms that is tetrahedrally coordinated by X atoms. The figure is adopted from [72].

The crystal structures of the three most often studied types, namely the 1111-type, the 122-type, and the 11-type are shown in Fig. 4.1. They consist of a superconducting layer that is identical for all of them and has the composition M_2X_2, where M is a metal atom and X is either a pnictogen $Pn =$ P, As, Sb or a chalcogen $Ch =$ S, Se, Te. At high temperatures the layers consist of a square lattice of M atoms tetrahedrally coordinated by X atoms that can undergo a orthorhombic distortion at low temperatures controlled by the actual doping. Depending on the type of superconductor the superconducting M_2X_2 layers are separated by layers of different thickness. Therefore, the 11-type is considered to be the most simple one among all the Fe-based superconductors, as it consists of FeCh layers only without any separating atoms.

The origin of superconductivity in the Fe-based materials still needs to be resolved. Already shortly after the discovery of superconductivity in LaFeAsO$_{1-x}$F$_y$ calculations indicated that conventional electron-phonon coupling is not sufficient to explain the high transition temperatures [73]. Several experiments show that the Fe-based superconductors are multi-band materials in which several gaps open in the superconducting state

Chapter 4. The FeCh system

(for a summary of the experiments see e.g. [74]). However, after three years of research no consensus of an universal gap symmetry has been reached. Already small changes in the electronic structure can lead to a significant diversity in the superconducting gap structure like gaps with nodes in some materials or nodeless gaps in others [24]. Nevertheless, the general symmetry class of the order parameter of most or even all of the Fe-based superconductors may be the same s-wave type. Some recent studies, however, also propose a d-wave symmetry, where nodes are on the hole pockets due to symmetry reasons and involve a sign change after 90° rotation [24, 75, 76]. But nevertheless the symmetry properties are distinct from the gap structure. Gaps with the same symmetry might have a different structure: The isotropic and fully gaped s-wave states s_{++} [77] and $s_{\pm}$ [78, 79] differ only by a relative phase shift π in the latter case between the hole and electron pockets. There is one more s case, the so-called nodal $s_{\pm}$ case [24, 75, 80], in which the gap is shown to vanish at certain points on the electron pockets but the overall sign is still opposite to the hole pockets. However, several publications proposing theoretical models for superconductivity and magnetism in Fe-based superconductors argue that within them the observed coexistence of a spin density wave and superconductivity is easily possible in the sign changing $s_{\pm}$ state whereas in the s_{++} state its only possible in a very narrow range of parameters [24, 81, 82, 83].

But even if the pairing symmetry would be known it does not itself imply a specific mechanism leading to superconductivity, although it appears to rule out electron-phonon interaction [84, 85, 86]. But this conclusion could be a bit premature, because a significant Fe-isotope effect on T_c was observed [87, 88, 89, 90, 91]. This indicates that superconductivity is somehow coupled to an electron-phonon interaction. Furthermore, it seems that Fermi surface nesting plays a major role in determining the gap properties of the Fe-based systems [92].

It has been shown by inelastic neutron scattering studies that spin fluctuations occur above T_c in the Fe-based superconductors, whereas all FeAs compounds show a diffraction peak at the same wave vector $Q_\text{nesting} = (\frac{1}{2}, \frac{1}{2})$ r.l.u. (reciprocal lattice units) (see e. g. [93] and references therein).

This is the nesting wave vector between the electron and hole Fermi surface. But also the FeCh system exhibits magnetic fluctuations in the superconducting samples above T_c along the same nesting wave vector Q_nesting [93, 94, 95, 96]. This, on the other hand, seems to point towards an antiferromagnetic spin fluctuation model for the pairing mechanism which would be consistent with the fully gapped $s_\pm$ pairing symmetry [97].
However, it appears that the suggested nodal states "may be impurity or 'accidental' effects rather than intrinsic features of the Fe-based superconductors" [93].
Attention has shifted from the 1111 to the 122 and especially to the 11 compounds, even though they have a lower T_c. The particular interest in the 11-type has several reasons like e.g. the simplicity of the system or mainly the availability of large single crystals allowing a more definite characterization of the physical properties [66].

4.2 Synthesis of FeSe$_{1-x}$

Two different routes to synthesize superconducting FeSe$_{1-x}$ were proposed. The first one is the so-called "low-temperature synthesis" (LTS) that uses Se and Fe powders as starting materials in a temperature range of $400 - 700\,°\text{C}$ [64]. Therefore, the Fe and Se powders were mixed and cold-pressed and afterwards the mixtures were sealed in quartz ampules. The FeSe$_{1-x}$ was then prepared in a solid state reaction by heating it up to $700\,°\text{C}$, slow cooling to room temperature, followed by intermediate grinding in an inert atmosphere and annealing at $400\,°\text{C}$ [64, 68, 98].
The second method is the "high-temperature synthesis" (HTS) starting from Fe pieces and Se shots that were first heated to $750\,°\text{C}$. Then the material is molten at $1075\,°\text{C}$ followed by a fast decrease in temperature down to $420\,°\text{C}$ and quenched afterwards. To complete the HTS the samples were annealed again in a new ampule at temperatures between $300\,°\text{C}$ and $500\,°\text{C}$ followed again by quenching [99].
Surprisingly, in the beginning the FeSe$_{1-x}$ samples synthesized by the LTS and HTS techniques were found to be rather different despite of

Chapter 4. The FeCh system

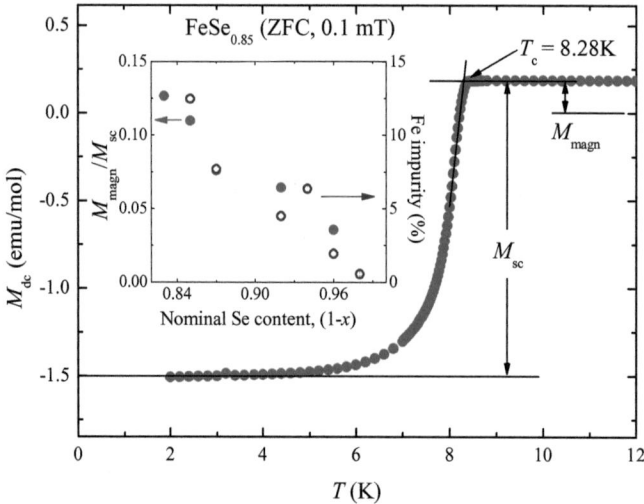

Figure 4.2: Temperature dependence of the dc magnetization M_{dc} of FeSe$_{0.85}$ (after zero field cooling in $\mu_0 H = 0.1\,\text{mT}$). The superconducting transition temperature T_c, the superconducting M_{SC} and the magnetic M_{magn} responses of the sample are determined as shown in the figure. The inset shows the dependencies of $M_{\text{magn}}/M_{\text{SC}}$ and Fe impurity concentration as a function of the nominal Se content. After [98].

$T_c \simeq 8\,\text{K}$ for both methods. Whereas in the LTS samples superconductivity was found first in a rather extended range of nominal Se content up to $x = 0.18$ in the HTS samples superconductivity was detected only in a very narrow region in the phase diagram ($0.01 \geq x \geq 0.025$). In addition, McQueen et al. reported that below 300 °C the tetragonal - and superconducting FeSe$_{1-x}$ converts into a hexagonal not superconducting NiAs-type phase [99]. Therefore, the samples were quenched from temperatures above 300 °C in the HTS. Whereas no special care for fast cooling of LTS samples needs to be taken [64, 98, 100]. In order to resolve these controversies, Pomjakushina et al. performed comparative studies of FeSe$_{1-x}$ synthesized by both the LTS and HTS method [98]. The magnetization measurements showed a $T_c \simeq 8\,\mathrm{K}$ for all FeSe$_{1-x}$ samples prepared both by LTS and HTS (the superconducting transition of one LTS sample is displayed in Fig. 4.2). The neutron powder diffraction (NPD) measurements on the same samples revealed that there is only one superconducting stoichiometry, namely FeSe$_{0.980(3)}$ [98]. Any deviation of the nominal stoichiometry from this one leads to an increase of secondary impurity phases. This is confirmed by the magnetization measurements above T_c that show a magnetization $M_\mathrm{magn} = 0$ only for the sample with the nominal composition FeSe$_{0.98}$. As seen in the inset of Fig. 4.2, the more the nominal stoichiometry deviates from $x = 0.02$ the larger is M_magn above T_c. Additionally, it is seen that M_magn scales with the amount of impurities detected by NPD.

By studying the temperature dependence of the crystal structure NPD revealed a second-order phase transition from a tetragonal to an orthorhombic phase on cooling at $T \sim 100\,\mathrm{K}$. Yet the Fe-Se-Fe bond angles become different in the low-temperature phase, the Se-height above the Fe-plane stays constant [98, 100].

4.3 Hydrostatic pressure effect

The superconducting transition temperature of FeSe$_{1-x}$ was found to be only $T_c \simeq 8\,\mathrm{K}$ [64]. Shortly after, pressure dependent studies of the electronic and magnetic phase diagram revealed one of the largest pressure effects on T_c known at present. The transition temperature reaches values of $T_c \approx 37\,\mathrm{K}$ at $p \approx 9\,\mathrm{GPa}$, demonstrating that FeSe$_{1-x}$ is in fact a high temperature superconductor [68, 101]. Furthermore, it was discovered that superconductivity disappears at very high pressures ($p \geq 9\,\mathrm{GPa}$), which is connected to a structural phase transition to a more densely packed hexagonal phase [68].

A more detailed investigation of $T_c(p)$ with finer pressure steps at low pressures up to $\sim 2.5\,\mathrm{GPa}$ revealed a nonlinear increase. The pressure dependence of T_c exhibits a local maximum at $p \simeq 0.8\,\mathrm{GPa}$ followed by a decrease of T_c to a local minimum at $p \simeq 1.2\,\mathrm{GPa}$ [102, 103, 104]. At higher pressures ($p \geq 1.2\,\mathrm{GPa}$) T_c increases again (see Fig. 4.3). It

Chapter 4. The FeCh system

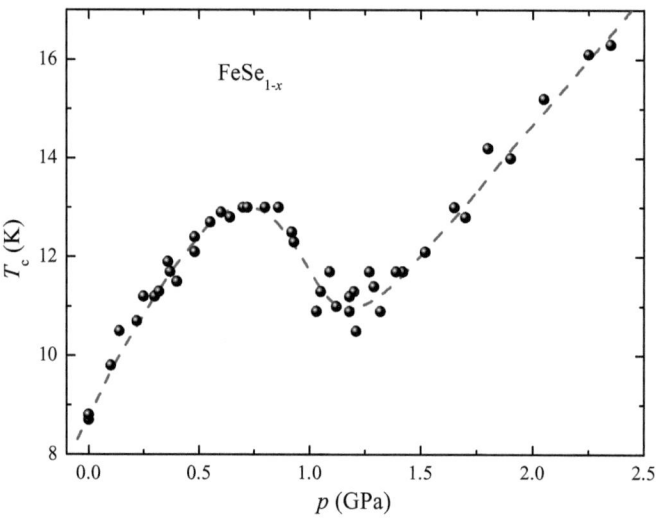

Figure 4.3: Dependence of the superconducting transition temperature T_c on pressure p of FeSe$_{1-x}$ [102, 105, 106]. Clearly, T_c increases almost linearly at low pressures up to $p \simeq 0.8\,\mathrm{PGa}$ where a local maximum is observed. A further increase of the pressure leads to a decrease of T_c to a local minimum at $p \simeq 1.2\,\mathrm{GPa}$ followed by an anew increase of T_c to the highest pressure investigated. The red dotted line is a guide to the eyes. After [102, 105].

was shown that in the region where T_c decreases static magnetic order develops in FeSe$_{1-x}$ and competes with superconductivity [102]. But as soon as the local minimum is reached and T_c increases again, superconductivity and the fully developed magnetism coexist within the whole sample volume.

The dc and ac measurements at ambient pressure without the pressure cell, performed in a SQUID magnetometer and a PPMS ac susceptometer showed that the samples are bulk superconductors with a susceptibility of $\chi \simeq -1.3$. By assuming each grain has a shape of a sphere and taking the corresponding demagnetization factor of $N = 1/3$ into account, leads to an ideal response of a superconductor of $\chi \simeq -1$. The measurements

4.3. Hydrostatic pressure effect

of the pressure dependence of T_c were performed in a double wall pressure cell especially designed for µSR measurements made from MP35 alloy. A frequency of $\nu_{ac} = 94\,\text{Hz}$ and an ac amplitude of $\mu_0 H_{ac} \approx 0.1\,\text{mT}$ was used for all measurements. It is well known that weak links can occur and that it can be checked for them by frequency and amplitude dependent measurements. None of the measurements gave an indication for weak links [105]. To determine the ac response under pressure reliably, the coils were directly wound on the pressure cell and the signal was normalized to the one obtained in a SQUID magnetometer. Based on these measurements the conclusion is that FeSe$_{1-x}$ is a bulk superconductor for all pressures investigated with a superconducting volume fraction of $\simeq 100\%$, as the ac response was constant up to the highest pressure investigated.

The superfluid density response was studied in transverse field (TF, $\mu_0 H = 0.01$ T) µSR experiments via the measurement of the in-plane magnetic penetration depth λ_{ab}. The superfluid density ρ_s can be expressed in terms of $\rho_s = n_s/m^*_{ab} \propto \lambda_{ab}^{-2}$ (n_s is the charge carrier density and m^*_{ab} is the carrier mass).

In a powder sample the magnetic penetration depth λ can be extracted from the Gaussian muon-spin depolarization rate $\sigma_{sc}(T) \propto 1/\lambda^2(T)$, which probes the second moment of the magnetic field distribution in the superconductor in the mixed state. σ_{sc} can be converted into λ_{ab} via [39]:

$$\sigma_{sc}^2/\gamma_\mu^2 = 0.00126\,\Phi_0^2/\lambda_{ab}^4, \qquad (4.1)$$

where $\Phi_0 = 2.068 \cdot 10^{-15}$ Wb is the magnetic flux quantum and $\gamma_\mu = 2\pi\,135.5$ MHz/T is the muon gyromagnetic ratio.

The measured $\lambda_{ab}^{-2}(T)$ of FeSe$_{0.94}$ at ambient pressure is shown in Fig. 4.4. Note that the determination of $\lambda_{ab}(T)$ was possible only for $p \lesssim 0.9$ GPa. At higher pressures the emergence of static, temperature dependent magnetism does not allow to obtain $\lambda_{ab}^{-2}(T)$ with a sufficient precision [102]. The experimentally obtained $\lambda_{ab}^{-2}(T)$ was analyzed within the framework of the so-called two-band weak coupling γ-model that accounts for the interband and intraband coupling, the partial densities of states at the Fermi level, and the Fermi velocities [109]. It allows to evaluate

Chapter 4. The FeCh system

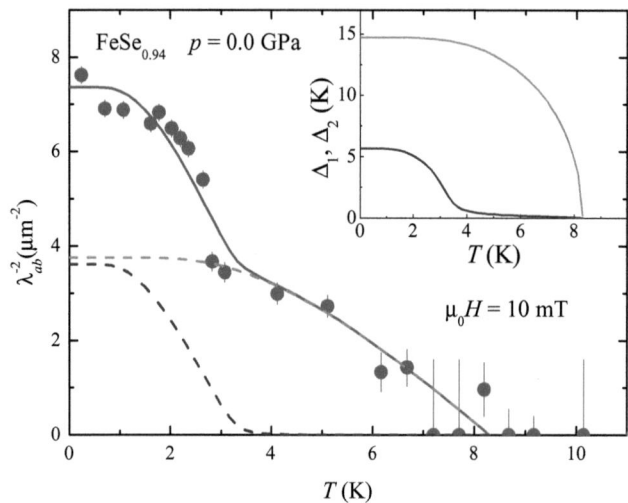

Figure 4.4: Temperature dependence of $\lambda_{ab}^{-2} \propto \rho_s = n_s/m*$ of FeSe$_{0.94}$ measured at ambient pressure in a CuBe pressure cell. Approximately 50% of the signal originates from the sample. The solid and dashed lines are the theoretical curves obtained within the framework of the γ-model [107, 108, 109]. The inset show the corresponding temperature dependencies of the large (Δ_1) and the small (Δ_2) gap. After [110].

self-consistently the temperature dependence of both superconducting energy gaps and the superfluid density. The theoretical superfluid density, calculated within the framework of the weak-coupling γ-model is shown as a solid red line in Fig. 4.4. The green and blue dashed lines represent the contributions of the large and the small gap to the superfluid density $\lambda_{ab,1}^{-2}(T)$ and $\lambda_{ab,2}^{-2}(T)$, respectively. The dependence of the superconducting gaps Δ_1 and Δ_2 on T are shown in the inset of Fig. 4.4. The dependence of $\lambda_{ab}^{-2}(0)$, $\lambda_{ab,1}^{-2}(0)$ and $\lambda_{ab,2}^{-2}(0)$, and $\Delta_1(0)$ and $\Delta_2(0)$ on T_c (or pressure) are shown in Figs. 4.5 (a), (b), and (c). In Fig. 4.5 (a) the tendency of an increasing superfluid density $\rho_s \propto \lambda_{ab}^{-2}$ with increasing T_c is seen, following an Uemura relation which has been established for various Fe-based superconductors [111]. Figures 4.5 (b) and

4.3. Hydrostatic pressure effect

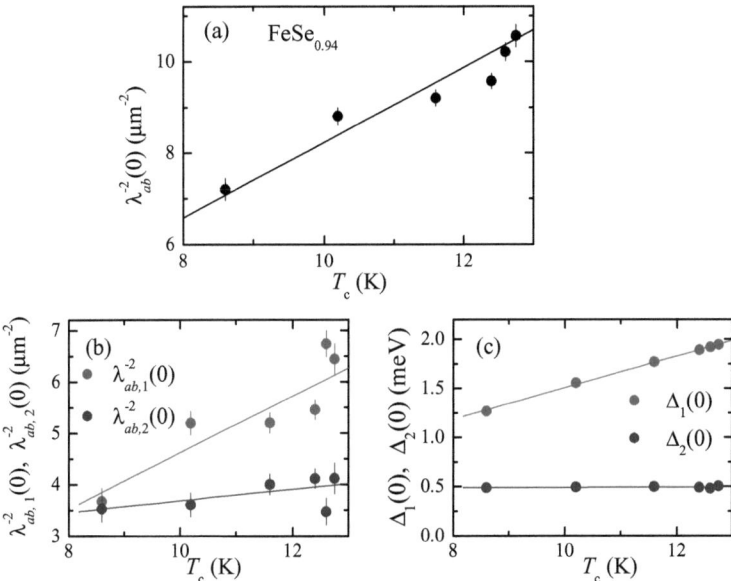

Figure 4.5: (a) Dependence of the superfluid density $\lambda_{ab}^{-2} \propto \rho_s = n_s/m^*$ on the superconducting transition temperature T_c that depends almost linear on pressure in the region shown. (b) Relative contribution of the superfluid density of the large gap $\lambda_{ab,1}^{-2}(0)$ and the small gap $\lambda_{ab,2}^{-2}(0)$ to the total superfluid density $\lambda_{ab}^{-2}(0)$ as a function of T_c. (c) Dependence of the zero-temperature values of the large gap $\Delta_1(0)$ and the small gap $\Delta_2(0)$ on T_c. After [110].

(c) show that the electronic bands are affected very differently by the pressure. Whereas the large gap $\Delta_1(0)$ and the corresponding $\lambda_{ab,1}^{-2}(0)$ show a strong dependence on T_c and increase almost linear, the small gap $\Delta_2(0)$ and the corresponding $\lambda_{ab,2}^{-2}(0)$ are hardly affected. Taking this together with the interband coupling constant that is estimated to be of the order of $\sim 10^{-4}$ or even less, one may conclude that the pressure effect on T_c and λ_{ab}^{-2} is mainly determined by the bands exhibiting the large superconducting gap.

29

Chapter 4. The FeCh system

The magnetic response of FeSe$_{1-x}$ under pressure was studied in ZF μSR experiments. Three different pressure regions with different magnetic and superconducting states emerge in the sample. In the low-pressure region ($0 \leq p \lesssim 0.8$ GPa) in which T_c increases almost linear with p (see Fig. 4.3) no static magnetic ordering was observed to the lowest temperatures investigated ($T \simeq 0.3$ K).

For pressures above $\simeq 0.8$ GPa a spontaneous muon spin precession is observed indicating that static long range magnetic order in the μSR time scale is present below the Néel temperature T_N. The oscillation is found to be best described by a zeroth-order spherical Bessel function, archetypical for incommensurate magnetic order [112]. However, as soon as superconductivity emerges the magnetic order is suppressed again, which is reflected in the temperature dependences of the internal magnetic field B_{int} that is proportional to the order parameter (Fig. 4.6(a)) and in the volume fraction (Fig. 4.6(b)) for different pressures. With increasing pressure an increase of both the volume fraction and the order parameter is observed. Furthermore, magnetism is less suppressed in the superconducting state.

At pressures higher than $p \geq 1.2$ GPa the magnetic volume fraction is close to 100% and stays constant even in the superconducting state. Also B_{int} does not decrease in the superconducting state any more, and the magnetic order becomes commensurate, as the μSR time spectra are best described by a cosine with zero initial phase. Additionally, with increasing pressure B_{int} increases further to the highest pressure investigated [102].

In an earlier study Medvedev et al. did not observe the occurrence of magnetic order in FeSe$_{1-x}$ under pressure by means of Mössbauer spectroscopy [101]. As already described in Section 4.2 two different methods HTS and LTS for the sample preparation are used so far. Medvedev et al. prepared the samples following the HTS method, whereas the μSR studies were done on samples prepared by the LTS method. To exclude that this discrepancy is a result of the different preparation procedures, samples for the μSR measurements were synthesized following exactly the HTS method proposed by McQueen et al. [99]. Both experiments

4.3. Hydrostatic pressure effect

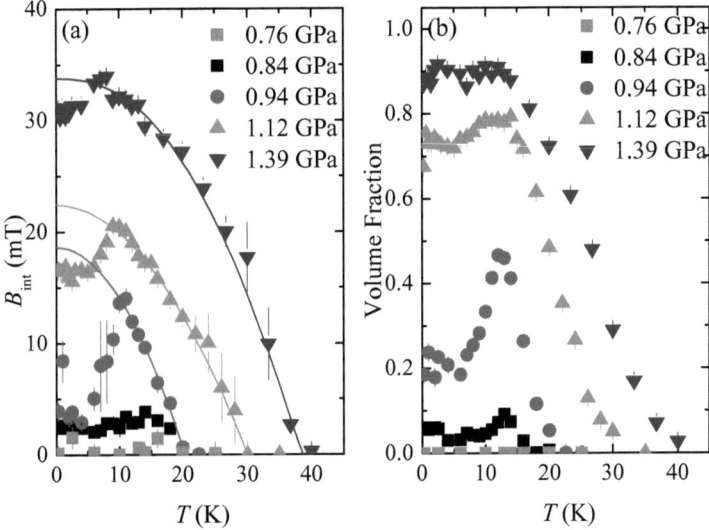

Figure 4.6: Temperature dependence of (a) the internal magnetic field at the muon stopping site B_{int} and (b) the magnetic volume fraction of FeSe$_{1-x}$ for various pressures. The solid lines in (a) are a fit of $B_{\text{int}}(T)$ in the region $T_c \leq T \leq T_N$ to $B_{\text{int}}(T) = B_{\text{int}}(0)[1 - (T/T_N)^\alpha]^\beta$ (α and β are power exponents). After [102].

lead to the same result allowing the conclusion that in the pressure and temperature range investigated by Medvedev et al. [101] no magnetic order appears [105, 106].

Up to now it is not clear what kind of magnetic order develops in FeSe$_{1-x}$ under pressure. Thus calculations of the muon stopping sites at different pressures combined with a symmetry analysis were performed to check for possible magnetic structures that were proposed. The muon stopping site calculations, using a modified Thomas-Fermi approach [113], revealed only one possible muon stopping site. It is located on the line connecting the Se-Se ions along the c-direction.

The studies of the crystal structure under pressure revealed a stronger reduction of the c-axis than of the a- and b-axis. This results in an increase of the Fe-Se-Fe bond angle that in accordance with the semi-empirical Goodenough-Kanamori rules can be interpreted as a tendency to antiferromagnetic exchange [114, 115, 116]. Knowing that a very small variation of the Fe-As-Fe bond angle gives rise to a drastic change of the magnetic exchange integral in the RFeAsO compounds [117]. This leads to the occurrence of a ferromagnetic type of order along the a-axis and antiferromagnetic along the b-axis in FeSe$_{1-x}$ under pressure. The minimal model accounting for this feature includes a doubling of the primitive cell along the b-direction with magnetic propagation vectors $K_I = (0; \pi/\tau_y; \pi/2\tau_z)$ or $K_{II} = (0; \pi/\tau_y; 0)$. However, other simpler possible magnetic vectors such as $K_0 = (0; 0; 0)$ and $K_{III} = (0; 0; \pi/2\tau_z)$ should also be considered.

The calculations of the magnitude and symmetry of the dipole fields of the Fe subsystem at the muon site for different pressures yield an increase of the internal field at the muon stopping site. This was observed in the experiments (see Fig. 4.6) only for the K_I and K_{II} translation symmetries, whereas for the K_0 and K_{III} translation symmetries the internal field decreases. Comparison of both possible magnetic structures K_I and K_{II} with the experimental data leads to magnetic fields along the z-coordinate of the size $B_z(K_I) = 354.6 \cdot m_y(K_I)$ and $B_z(K_{II}) = 334.3 \cdot m_y(K_{II})$, respectively. Both magnetic structures K_I and K_{II} provide similar and very low values of the magnetic moment per Fe atom $\mu \approx 0.2\mu_B$ at the lowest temperatures and at the highest pressure investigated in this study ($p \simeq 2.4$ GPa) [105].

In order to test the proposed models neutron powder diffraction experiments were performed at $p = 4.4(5)$ GPa. To obtain the position of the possible magnetic peak the background of the diffraction patterns taken at 5 and 150 K were normalized to each other and then subtracted. However, no difference peak of magnetic origin was observed. Simulations of the proposed magnetic structures with the expected magnetic moments at $p = 4.4(5)$ GPa resulted in magnetic reflections much smaller than the background signal [105].

The electronic phase diagram studied up to $p \simeq 2.5$ GPa is shown in Fig. 4.7. At low pressures below $p \leq 0.8$ GPa the samples are only

4.3. Hydrostatic pressure effect

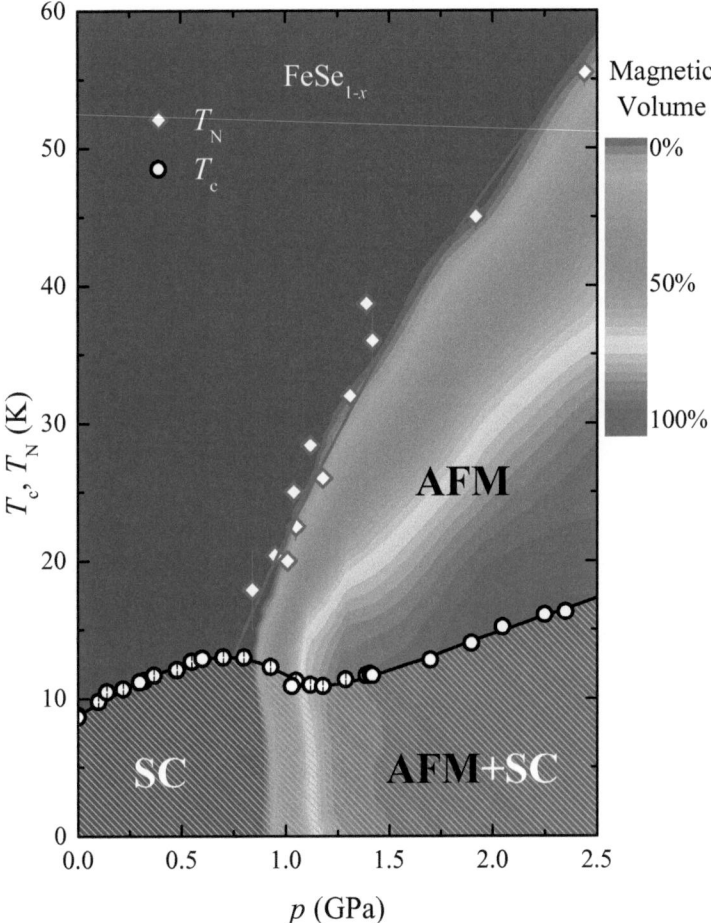

Figure 4.7: Pressure dependence of the superconducting transition temperature T_c, the magnetic ordering temperature T_N, and the superconducting and magnetic volume fractions (the superconducting volume is 100% for all pressures investigated) obtained by ac susceptibility and muon spin rotation experiments for $FeSe_{1-x}$. The lines marking the pressure dependence of T_c and T_N are guides to the eyes. SC and AFM denote the superconducting and magnetic states of the samples, respectively. The nonmagnetic state is indicated by the blue area. After [102] and [105].

Chapter 4. The FeCh system

superconducting and show an almost linear increase of T_c from $\sim 8\,\text{K}$ at ambient pressure to $\sim 13\,\text{K}$ at $\simeq 0.8\,\text{GPa}$. At higher pressures static magnetic in the muon time scale order occurs below $T_N > T_c$ that first competes and coexists with superconductivity and then, for $p \gtrsim 1.2\,\text{GPa}$, coexists only with superconductivity. In the intermediate pressure range ($0.8 \leq p \leq 1.2\,\text{PGa}$) the competition is evident from two observations: First, as a function of pressure T_c is suppressed as soon as magnetic order appears, leading to the local maximum of T_c at $p \simeq 0.8\,\text{GPa}$ (see Fig. 4.3). However, the superconducting volume fraction stays 100% for all pressures investigated. Second, the magnetic order present above T_c is parially (or even fully) suppressed by the onset of superconductivity. This is evident in the decrease of the internal magnetic field B_{int} and the decrease of the magnetic volume fraction shown in Fig. 4.6 [102, 105]. Above $p \simeq 1.2\,\text{GPa}$ magnetism is fully established, and both T_N and the magnetic moment increase with increaing pressure.

The appearance of antiferromagnetic order has also been seen by NMR experiments [118]. An increase of $1/TT_1$ close to T_c is observed at low pressures ($p = 0$ and $0.7\,\text{GPa}$) indicating antiferromagnetic modes of spin fluctuations that are strongly enhanced towards T_c. This leads to the conclusion that $FeSe_{1-x}$ is in close proximity to a magnetic instability. At higher pressures ($p = 1.4$ and $2.2\,\text{GPa}$, where μSR observes static magnetic ordering) the $1/TT_1$ data reveal a broad hump significantly above T_c. Furthermore, the integrated intensity of the NMR signal begins to decrease at temperatures of $\sim 34\,\text{K}$ in $1.4\,\text{GPa}$ and $\sim 50\,\text{K}$ in $2.2\,\text{GPa}$ in excellent agreement with the μSR data. The disappearance of the NMR signal below a peak of $1/TT_1$ is a typical sign for a magnetic phase transition [118].

Keeping in mind that the superconducting volume fraction is 100% for all pressures measured and that the magnetic volume fraction reaches 100% at pressures above $1.2\,\text{GPa}$ indicates that both ground states coexist in the whole sample volume. The data does not show any signature for macroscopic phase separation into superconducting and magnetic regions bigger than a few nm size, as observed e.g. in $Ba_{1-x}K_xFe_2As_2$ [119] or $LaFeAsO_{1-x}F_x$ [120]. Moreover, no sublattice that could order magnetically is present while the superconducting FeAs layers are not magnetically ordered, as e.g. in Ce- or Sm1111 [121, 122]. All these findings point rather to an atomic coexistence of the order parameters as it is seen e.g. in FeTe$_{1-x}$Se$_x$ [123] or Ba(Fe$_{1-x}$Co$_x$)$_2$As$_2$ [124]. Furthermore, it seems that the two ground states stabilize each other with pressure, since T_c, T_N, and B_{int} are all increasing hand in hand with increasing pressure. A comparison with the newly discovered RFe$_{2-x}$Se$_2$ 245 system (R = K, Cs, Rb) where superconductivity and magnetism coexist and which has the same crystallographic structure as the 122 family, rises the question whether the magnetic order in FeSe$_{1-x}$ under pressure and the magnetic order in the 245 system are of similar origin [125, 126]. In the 245 system the superconducting transition temperatures reaches $T_c \simeq 32$ K and superconductivity seem to coexists with magnetism occuring at $T_N \approx 500$ K with a rather large magnetic moment of $3\mu_B$ per Fe atom [127].

Knowing that FeSe$_{1-x}$ is a two gap superconductor [110, 128] a possible scenario of an atomic coexistence of superconductivity and magnetism has recently been proposed by Vorontsov *et al.* [81, 83, 129] and Cvetkovic and Tesanovic [130]. They proposed a region where superconductivity and magnetic order can coexist. In this region the magnetic order can be commensurate only in a rather small parameter range where the Fermi surface nesting is not perfect, the bands are supposed to have an elliptical shape, and the chemical potential is supposed to shift.

4.4 Chemical pressure effect and role of Fe

Application of hydrostatic pressure is not the only possibility to increase T_c in the 11-system. Very shortly after superconductivity was found in α-FeSe$_{1-x}$ [64] at $T_c \simeq 8$ K it was realized that substitution of Se by the isovalent Te or S, so to say applying chemical pressure, leads to an increase of T_c up to $\simeq 14$ K at a substitution level of ≈ 50% Se by Te [66]. Furthermore, it was discovered that large single crystals of Fe$_y$Se$_x$Te$_{1-x}$ can be grown for $0 \leq x \leq 0.5$ [66].

The availability of large superconducting single crystals made it possible to study the anisotropic superconducting properties of $Fe_ySe_xTe_{1-x}$. They were grown by the Bridgman method by mixing the appropriate amounts of Fe, Se, and Te powders together. The mixed powder was then pressed into a rod that was put to a double wall ampule and evacuated. The ampule was placed into a vertical furnace with a temperature gradient and heated to $T = 1200\,°C$. The melt was kept for 4 hours at that temperature. Afterwards, the samples were cooled down to $750\,°C$ with a rate of $4\,°C/h$, followed by a fast cooling ($\sim 50\,°C/h$) to room temperature. The so-obtained crystals were easily cleaved from the as-grown crystal along the ab plane.

The superconducting properties of $Fe_ySe_xTe_{1-x}$ were studied by means of magnetization measurements on a piece of the crystal with the nominal stoichiometry $FeSe_{0.5}Te_{0.5}$ and a mass of about 200 mg. A sharp transition to the superconducting state at $T_c = 14.6\,K$ was detected. However, the large difference between the field cooled and zero field cooled magnetization is a signature, that strong vortex pinning is present. This is in agreement with the weakly field dependent and pronounced critical current density of the order of $10^7\,A/m^2$ [111].

The temperature dependence of the lower critical field H_{c1} was determined from magnetization measurements with the field applied parallel and perpendicular to the ab-plane yielding the zero temperature values of $H_{c1}^{\|ab}(0) = 2.0(2)\,mT$ and $H_{c1}^{\|c}(0) = 4.5(3)\,mT$. From these values the zero temperature magnetic penetration depths were estimated to $\lambda_{ab}(0) \approx 460(100)\,nm$ and $\lambda_c(0) \approx 1100(300)\,nm$. These values are in good agreement with the penetration depths obtained by μSR [111].

The temperature dependence of the magnetic penetration depth in both crystallographic directions was determined also by μSR. This was done by applying the field once parallel and once perpendicular to the ab-plane, but always perpendicular to the muon spin. The so obtained anisotropic μSR line shapes were analyzed with the help of the Ginzburg-Landau model [38]. The temperature dependence of the magnetic penetration depth is shown in Fig. 4.8. It was best described by a $s+s$-wave model within the framework of the so-called α-model [131] that assumes that λ^{-2} is a linear combination of two terms. The zero tem-

4.4. Chemical pressure effect and role of Fe

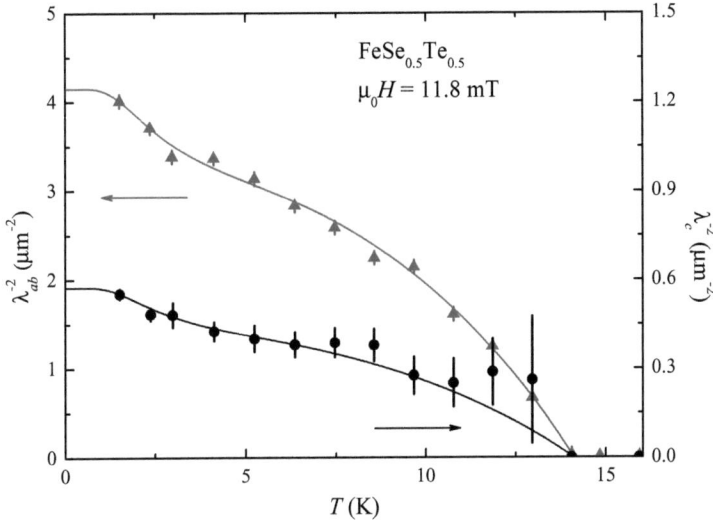

Figure 4.8: Temperature dependence of the penetration depth components λ_{ab} and λ_c of single crystal FeSe$_{0.5}$Te$_{0.5}$. The solid lines represend fits to the α-model. After [111].

perature values of the magnetic penetration depth were determined to be $\lambda_{ab}(0) = 491(8)$ nm and $\lambda_c(0) = 1320(14)$ nm [111], consistent with previous μSR results [132].

From the temperature dependence of λ_{ab} and λ_c the gap to T_c ratios $2\Delta_S^0/k_BT_c = 0.84(4)$ and $2\Delta_L^0/k_BT_c = 4.3(1)$ were obtained. These values are in fair agreement with μSR results of FeSe$_{0.5}$Te$_{0.5}$ [132] and the results obtained by other methods [135, 136]. They are also very close to what is reported for isotstructural FeSe$_{1-x}$ (see above and Refs. [110] and [128]). Furthermore, Evtushinski et al. pointed out that most of the Fe-based superconductors exhibit two-gap superconductivity with a large gap of $2\Delta_L^0/k_BT_c = 7(2)$ and a small gap of $2\Delta_S^0/k_BT_c = 2.5(1.5)$ [74]. The corresponding values of FeSe$_{0.5}$Te$_{0.5}$ are at the lower limit of these estimations.

Chapter 4. The FeCh system

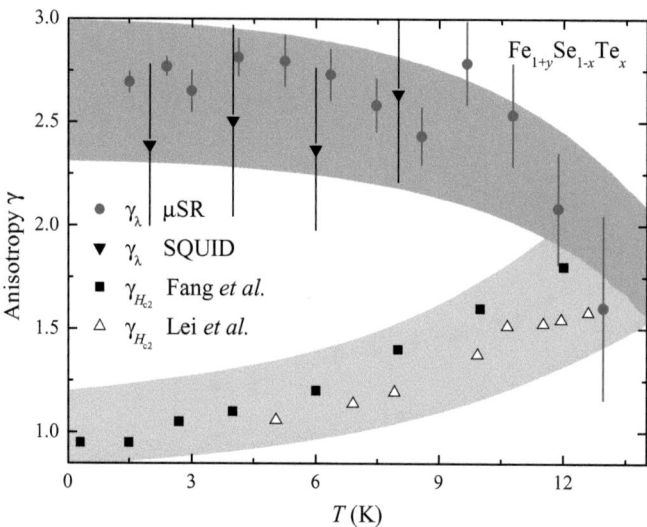

Figure 4.9: Temperature dependence of the magnetic penetration depth anisotropy parameter $\gamma_\lambda = \lambda_c/\lambda_{ab}$ derived from μSR and SQUID measurements on single crystalline FeSe$_{0.5}$Te$_{0.5}$ and the upper critical field H_{c2}-anisotropy parameter $\gamma_{H_{c2}} = H_{c2}^{\|ab}/H_{c2}^{\perp ab}$ obtained from resistivity measurements [133, 134]. The lines are guides to the eyes. After [111].

Moreover, from the T-dependence of λ_{ab} and λ_c obtained from μSR the magnetic penetration depth anisotropy parameter γ_λ was extracted. It is within the error the same as the one deduced from H_{c1} measurements by a SQUID (see Fig. 4.9). Both techniques yield a T-dependent γ_λ that increases with decreasing temperature from 1.6 at T_c to 2.6 at $T = 1.6$ K. While γ_λ increases with decreasing temperature, the anisotropy parameter of the upper critical field $\gamma_{H_{c2}}$ determined by resistivity measurements decreases with decreasing temperature [133, 134]. The observed T-dependence of the anisotropy parameters γ_λ and $\gamma_{H_{c2}}$ is similar to the one seen earlier by Weyeneth *et al.* in the 1111 family [137, 138]. However, superconductivity in FeSe$_{0.5}$Te$_{0.5}$ is much more isotropic suggesting that the direct electronic coupling of the Fe$_2$Se$_2$ layers in the

4.4. Chemical pressure effect and role of Fe

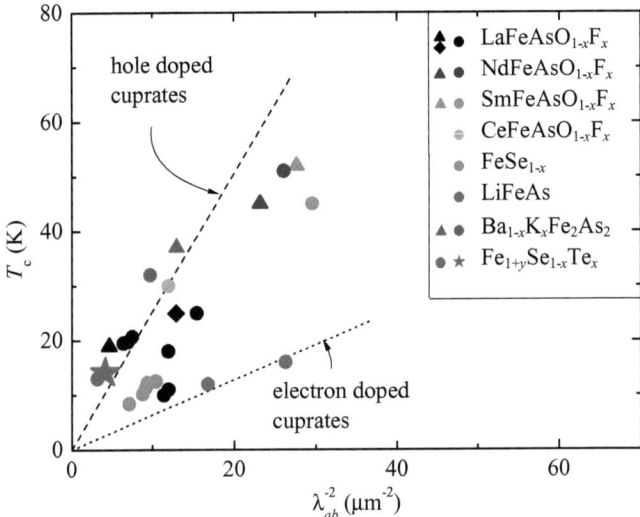

Figure 4.10: Uemura plot for a selection of Fe-based high temperature superconductors. The Uemura relation observed for underdoped cuprates is included for comparison (dashed line: hole doped, dotted line: electron doped cuprates). After [111] and references therein.

11 family is more effective than the coupling through the LnO layers in the 1111 Fe-based systems [111]. Such an anisotropic behavior was also observed in the known two-gap superconductor MgB_2, however, there γ_λ and $\gamma_{H_{c2}}$ have reversed sign: γ_λ decreases with decreasing temperature while $\gamma_{H_{c2}}$ increases.

The value of $\lambda_{ab}^{-2}(0)$ for $FeSe_{0.5}Te_{0.5}$ extracted from μSR data as well as the values of $\lambda_{ab}^{-2}(0)$ obtained for various Fe-based superconductors fall on the Uemura plot [140] (see Fig. 4.10) within the limits of hole-doped and electron-doped cuprates [141]. This suggests that the pairing mechanism in the Fe-based superconductors is unconventional as in the cuprates.

The maximum T_c in $Fe_y Se_x Te_{1-x}$ is reached at a substitution level of $\sim 50\%$. A further increase of $1-x$ leads to the appearance of magnetic

Chapter 4. The FeCh system

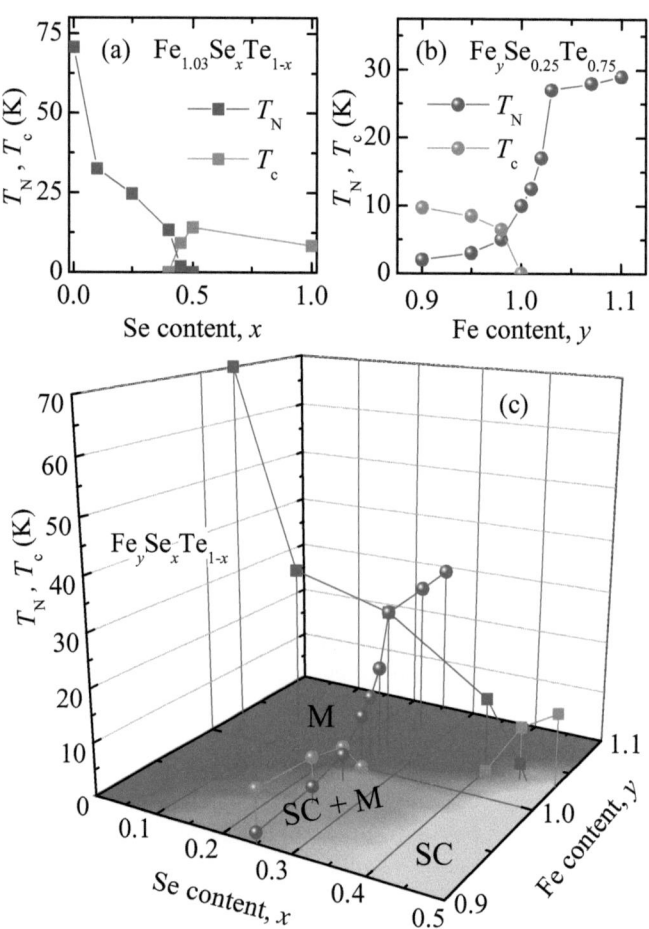

Figure 4.11: (a) Phase diagram of $Fe_{1.03}Se_xTe_{1-x}$ as a function of x. The datum of $FeSe_{1-x}$ is taken from Ref. [128]. (b) Variation in T_c and T_N in $Fe_ySe_{0.25}Te_{0.75}$ as a function of y. (c) The three dimensional phase diagram of T_c and T_N of $Fe_ySe_xTe_{1-x}$ as a function of x and y. M and SC denotes the magnetic and superconducting phases, respectively. With increasing amount of Se superconductivity is established, whereas it strongly depends on the amount of Fe in the system. After [123] and [139].

4.4. Chemical pressure effect and role of Fe

order in the system, whereas superconductivity and magnetism coexist on atomic length scale in the region of $0.55 \lesssim 1 - x \lesssim 0.6$. Above $(1-x \gtrsim 0.6)$ only incommensurate magnetic order is present with traces of superconductivity [123]. With increasing amount of Te in the system T_N is increasing from $T_N \simeq 3\,\text{K}$ at $1 - x \simeq 0.45$ to $33\,\text{K}$ at $1 - x \simeq 0.9$. For $1 - x \gtrsim 0.9$ no traces of superconductivity are detected in the system and the magnetic order changes to commensurate long range antiferromagnetic order. In pure FeTe $T_N = 70\,\text{K}$ and coincides with a structural transition making the phase transition first order like with a hysteretic behavior [123]. The corresponding electronic phase diagram is shown in Fig. 4.11 (a).

Interestingly, the superconducting and magnetic properties of $Fe_y Se_x Te_{1-x}$ not only depend on the Se:Te ratio x but also strongly on the Fe content y [142]. This was studied systematically by keeping x constant to $x = 0.25$ and varying y in the range $0.9 \leq y \leq 1.1$ [139]. As can be seen in the phase diagram in Fig. 4.11 (b), for $y \lesssim 1$ (low Fe content region) bulk superconductivity is established and coexists with incommensurate magnetism. In this Fe deficient region, the magnetic correlations of the Fe magnetic moments are more short ranged as compared to the region of excess Fe and thus magnetic order is less well coordinated. Hence, only the reduction of the Fe content results in a reduction of the magnetic correlations and the system becomes superconducting. However, it seems unlikely that the excess Fe atoms acts as isolated magnetic moments that destroy superconductivity. They may act as magnetic electron donors that suppress superconductivity and induce weakly localized electronic states [139].

Combining both the phase diagram of $Fe_{1.03}Se_x Te_{1-x}$ and the one of $Fe_y Se_{0.25} Te_{0.75}$ leads to a tentative three-dimensional phase diagram shown in Fig. 4.11 (c). It is seen that $Fe_y Te$ is always antiferromagnetically ordered [143, 144]. Upon substituting Te by Se the order becomes weaker and superconductivity occurs and finally the system is a pure bulk superconductor. The superconducting and magnetic behavior of $Fe_y Se_x Te_{1-x}$ can therefore not only be tuned by substituting Te with Se, but also by adjusting the Fe content. It is suggested that superconductivity has to be of multi-band nature, where different doping channels may be involved [111].

4.5 Related publications to Chapter 4

4.5.1 Paper I: Synthesis, crystal structure, and chemical stability of the superconductor $FeSe_{1-x}$

This work is published in:

E. Pomjakushina, K. Conder, V. Pomjakushin, M. Bendele, and R. Khasanov, *Synthesis, crystal structure, and chemical stability of the superconductor $FeSe_{1-x}$*, Phys. Rev. B **80**, 024517 (2009).

Abstract:

We report on a comparative study of the crystal structure and the magnetic properties of $FeSe_{1-x}$ ($x = 0.0 - 0.15$) superconducting samples by neutron powder-diffraction and magnetization measurements. The samples were synthesized by two different methods: a low-temperature one using powders as a starting material at $T \simeq 700\,°C$ and a high-temperature method using solid pieces of Fe and Se at $T \simeq 1075\,°C$. The effect of a starting (nominal) stoichiometry on the phase purity of the obtained samples, the superconducting transition temperature T_c, as well as the chemical stability of $FeSe_{1-x}$ at ambient conditions were investigated. It was found that in the Fe-Se system, a stable phase exhibiting superconductivity at $T_c \simeq 8\,K$ exists in a narrow range of selenium concentration ($FeSe_{0.974 \pm 0.005}$).

URL: http://link.aps.org/doi/10.1103/PhysRevB.80.024517
DOI: 10.1103/PhysRevB.80.024517
PACS: 74.70.-b, 74.72.-h, 61.05.fm, 74.25.Ha
Copyright 2009 by The American Physical Society.

4.5.2 Paper II: Pressure Induced Static Magnetic Order in Superconducting FeSe$_{1-x}$

This work is published in:

M. Bendele, A. Amato, K. Conder, M. Elender, H. Keller, H.-H. Klauss, H. Luetkens, E. Pomjakushina, A. Raselli, and R. Khasanov, *Pressure Induced Static Magnetic Order in Superconducting FeSe$_{1-x}$*, Phys. Rev. Lett. **104**, 087003 (2010).

Abstract:

We report on a detailed investigation of the electronic phase diagram of FeSe$_{1-x}$ under pressures up to 1.4 GPa by means of ac magnetization and muon-spin rotation. At a pressure $\simeq 0.8$ GPa the nonmagnetic and superconducting FeSe$_{1-x}$ enters a region where static magnetic order is realized above T_c and bulk superconductivity coexists and competes on short length scales with the magnetic order below T_c. For even higher pressures an enhancement of both the magnetic and the superconducting transition temperatures as well as of the corresponding order parameters is observed. These exceptional properties make FeSe$_{1-x}$ to be one of the most interesting superconducting systems investigated extensively at present.

URL: http://link.aps.org/doi/10.1103/PhysRevLett.104.087003
DOI: 10.1103/PhysRevLett.104.087003
PACS: 74.70.-b, 74.25.Jb, 76.75.+i
Copyright 2010 by The American Physical Society.

4.5.3 Paper III: Evolution of Two-Gap Behavior of the Superconductor FeSe$_{1-x}$

This work is published in:

R. Khasanov, M. Bendele, A. Amato, K. Conder, H. Keller, H.-H. Klauss, H. Luetkens, and E. Pomjakushina, *Evolution of Two-Gap Behavior of the Superconductor FeSe$_{1-x}$*, Phys. Rev. Lett. **104**, 087004 (2010).

Abstract:

The superfluid density, ρ_s, of the iron chalcogenide superconductor, FeSe$_{1-x}$, was studied as a function of pressure by means of muon-spin rotation. The analysis of $\rho_s(T)$ within the two-gap scheme reveals that the effect on both, the transition temperature T_c and $\rho_s(0)$, is entirely determined by the band(s) where the large superconducting gap develops, while the band(s) with the small gap become practically unaffected.

URL: http://link.aps.org/doi/10.1103/PhysRevLett.104.087004
DOI: 10.1103/PhysRevLett.104.087004
PACS: 74.70.-b, 74.25.Jb, 74.62.Fj, 76.75.+i
Copyright 2010 by The American Physical Society.

4.5.4 Paper IV: Coexistence of incommensurate magnetism and superconductivity in $Fe_{1+y}Se_xTe_{1-x}$

This work is published in:

R. Khasanov, M. Bendele, A. Amato, P. Babkevich, A. T. Boothroyd, A. Cervellino, K. Conder, S. N. Gvasaliya, H. Keller, H.-H. Klauss, H. Luetkens, V. Pomjakushin, E. Pomjakushina, and B. Roessli, *Coexistence of incommensurate magnetism and superconductivity in $Fe_{1+y}Se_x$-Te_{1-x}*, Phys. Rev. B **80**, 140511 (2009).

Abstract:

We have studied the superconducting and magnetic properties of $Fe_{1+y}Se_xTe_{1-x}$ single crystals ($0 \leq x \leq 0.5$) by magnetic susceptibility, muon-spin rotation, and neutron diffraction. We find three regimes of behavior: (i) commensurate magnetic order for $x \lesssim 0.1$, (ii) bulk superconductivity for $x \sim 0.5$, and (iii) a range $x \approx 0.25 - 0.45$ in which superconductivity coexists with incommensurate magnetic order. The results are qualitatively consistent with two-band mean-field models in which itinerant magnetism and extended s-wave superconductivity are competing order parameters.

URL: http://link.aps.org/doi/10.1103/PhysRevB.80.140511
DOI: 10.1103/PhysRevB.80.140511
PACS: 74.70.-b, 74.25.Jb, 61.05.F-, 76.75.+i
Copyright 2009 by The American Physical Society.

4.5.5 Paper V: Tuning the superconducting and magnetic properties in Fe$_y$Se$_{0.25}$Te$_{0.75}$ by varying the Fe-content

This work is published in:

M. Bendele, P. Babkevich, S. Katrych, S. N. Gvasaliya, E. Pomjakushina, K. Conder, B. Roessli, A. T. Boothroyd, R. Khasanov, and H. Keller, *Tuning the superconducting and magnetic properties in Fe$_y$Se$_{0.25}$Te$_{0.75}$ by varying the Fe-content*, Phys. Rev. B **82**, 212504 (2010).

Abstract:

The superconducting and magnetic properties of Fe$_y$Se$_{0.25}$Te$_{0.75}$ single crystals ($0.9 \leq y \leq 1.1$) were studied by means of x-ray diffraction, superconducting quantum interference device magnetometry, muon-spin rotation, and elastic neutron diffraction. The samples with $y < 1$ exhibit coexistence of bulk superconductivity and incommensurate magnetism. The magnetic order remains incommensurate for $y \geq 1$ but with increasing Fe content superconductivity is suppressed and the magnetic correlation length increases. The results show that the superconducting and the magnetic properties of the Fe$_y$Se$_{1-x}$Te$_x$ can be tuned not only by varying the Se/Te ratio but also by changing the Fe content.

URL: http://link.aps.org/doi/10.1103/PhysRevB.82.212504
DOI: 10.1103/PhysRevB.82.212504
PACS: 74.70.Xa, 76.75.+i, 74.25.Dw, 78.70.Nx
Copyright 2010 by The American Physical Society.

Chapter 4. The FeCh system

4.5.6 Paper VI: Anisotropic superconducting properties of single-crystalline FeSe$_{0.5}$Te$_{0.5}$

This work is published in:

M. Bendele, S. Weyeneth, R. Puzniak, A. Maisuradze, E. Pomjakushina, K. Conder, V. Pomjakushin, H. Luetkens, S. Katrych, A. Wisniewski, R. Khasanov, and H. Keller, *Anisotropic superconducting properties of single-crystalline FeSe$_{0.5}$Te$_{0.5}$*, Phys. Rev. B **81**, 224520 (2010).

Abstract:

Iron-chalcogenide single crystals with the nominal composition FeSe$_{0.5}$Te$_{0.5}$ and a transition temperature of $T_c \simeq 14.6$ K were synthesized by the Bridgman method. The structural and anisotropic superconducting properties of those crystals were investigated by means of single crystal x-ray and neutron powder diffraction, superconducting quantum interference device and torque magnetometry, and muon-spin rotation (μSR). Room temperature neutron powder diffraction reveals that 95% of the crystal volume is of the same tetragonal structure as PbO. The structure refinement yields a stoichiometry of Fe$_{1.045}$Se$_{0.406}$Te$_{0.594}$. Additionally, a minor hexagonal Fe$_7$Se$_8$ impurity phase was identified. The magnetic penetration depth λ at zero temperature obtained by means of μSR was found to be $\lambda_{ab}(0) = 491(8)$ nm in the ab plane and $\lambda_c(0) = 1320(14)$ nm along the c axis. The zero-temperature value of the superfluid density $\rho_s(0) \propto \lambda^{-2}(0)$ obeys the empirical Uemura relation observed for various unconventional superconductors, including cuprates and iron pnictides. The temperature dependences of both λ_{ab} and λ_c are well described by a two-gap $s+s$-wave model with the zero-temperature gap values of $\Delta_S(0) = 0.51(3)$ meV and $\Delta_L(0) = 2.61(9)$ meV for the small and the large gap, respectively. The magnetic penetration depth anisotropy parameter $\gamma_\lambda(T) = \lambda_c(T)/\lambda_{ab}(T)$ increases with decreasing temperature, in agreement with $\gamma_\lambda(T)$ observed in the iron-pnictide superconductors.

4.5. Related publications to Chapter 4

URL: http://link.aps.org/doi/10.1103/PhysRevB.81.224520
DOI: 10.1103/PhysRevB.81.224520
PACS: 74.25.Ha, 74.25.Op, 74.70.Xa, 76.75.+i
Copyright 2010 by The American Physical Society.

5 Isotope effect

The first isotope studies were performed by Kamerlingh-Onnes in 1922 [145]. However, he had only the two natural lead isotopes with the isotope masses $M = 206$u and $M = 207.2$u to his hands. Within the measurement accuracy at that time no difference in the transition temperatures was found. Only almost 30 years later, after some progress in the nuclear physics area, it was possible to create isotopes with larger mass differences in sufficient amount. As a result of these achievements, both Maxwell [8] and Reynolds *et al.* [9] found independent from each other almost at the same time that T_c in mercury is an inverse function of M.

A few years after the first successful isotope experiments of Maxwell and Reynolds *et al.* [8, 9] the BCS-theory was developed [10, 11]. It is able to well explain the isotope effect. T_c in the weak coupling BCS theory is given by:

$$k_B T_c = 1.13 \hbar \omega_D \exp\left(-\frac{1}{N(0)V}\right) \qquad (5.1)$$

Here, ω_D is the Debye frequency that is proportional to $M^{-1/2}$. The product $N(0)V$ is the electron-phonon coupling constant, where V is the electron-phonon interaction strength and $N(0)$ the electron density of states at the Fermi level that are both independent from the isotope mass M.

Since $T_c \sim \omega_D \sim M^{-1/2}$ the isotope exponent α can be easily deduced from Eq. (5.1):

$$\alpha = -\frac{\delta \ln T_c}{\delta \ln M} = -\frac{\delta T_c/T_c}{\delta M/M} = \frac{1}{2} \qquad (5.2)$$

This relation is fulfilled for a series of conventional superconductors, such as Hg, Sn or In and directly demonstrates the influence of the electron-phonon interaction to the formation of the superconducting state. However, not all classical superconductors show an isotope exponent $\alpha = 0.5$. There are some exceptions that can be explained by the fact that the theory leading to Eq. (5.1) is based on a lot of simplifications, which, however, are valid for most of the conventional low temperature superconductors. By investigating V in Eq. (5.1) in more detail, it is possible to explain the deviations from the BCS value $\alpha = 0.5$. The coupling strength is basically the difference of the attracting electron-phonon interaction and the repulsing Coulomb interaction of the electrons. However, deviations from $\alpha = 0.5$ or even a full absence of the isotope effect cannot be taken as a prove that the electron-phonon coupling is not responsible for pairing.

Also in the newer high temperature superconductors, such as the fullerenes, MgB_2, or the cuprates, isotope effect studies were performed. In fact in all of them isotope effects were found: in the fullerenes $\alpha \simeq 0.3$ was obtained for substituting ^{12}C with ^{13}C which is in good agreement with the electron-phonon coupling by assuming that the intramolecular oscillations of the C_{60} molecules dominate the pairing effect [146, 147]. In MgB_2 the isotope effect for both Mg and B was investigated. Exchanging the isotopes between ^{11}B and ^{10}B lead to an exponent of $\simeq 0.3$ [148, 149, 150, 151] whereas the variation of the Mg isotopes from ^{24}Mg to ^{26}Mg result only to $\alpha \simeq 0.02$ [149, 150]. These findings reveal that only the electron-phonon coupling of the Boron ions play a major role in the pairing.

In the cuprate superconductors the situation was found to be more complicated; they show an unconventional oxygen isotope effect on T_c. The substitution of ^{16}O to ^{18}O yields to an isotope exponent α that is small, but positive for the optimally doped compounds and increases upon decreasing the doping level. At the phase boundary it even exceeds the BCS value of 0.5 [152, 153, 154]. Interestingly, the values of the copper isotope exponents resulting from the exchange ^{63}Cu to ^{65}Cu are close to the oxygen isotope ones [155, 152, 153, 154]. If the doping is further decreased and the superconducting dome is left also the other thermodynamic quantities as the Néel temperature T_N and the spin glass temperature T_{sg} of the cuprate superconductors show an isotope effect of which the exponent tends to diverge at the phase boundaries [156]. These substantial isotope effects in the cuprate superconductors indicate that lattice effects play a relevant role in the occurrence of superconductivity in the cuprates.

5.1 Isotope effect in the Fe-based superconductors

After the discovery of the Fe-based superconductors immediately the question arouse, whether they also show an isotope effect. Until now only a few studies are available. Liu et al. [87] showed an iron isotope effect exponent $\alpha \simeq 0.34$ in $SmFeAsO_{0.85}F_{0.15}$ with $T_c \simeq 42\,K$ by replacing ^{56}Fe with the isotope ^{54}Fe. In contrast Shirage et al. [88] published a vanishing small effect in the same family, namely $SmFeAsO_{1-y}$ with a value of $\alpha \simeq -0.018$ by substituting the heavier ^{57}Fe with the lighter ^{54}Fe. However, this sample had a substantially larger T_c of $\simeq 54\,K$ close to optimal doping. Comparing the two isotope effect exponents one could argue that the situation is similar to the cuprate superconductors where the oxygen isotope effect is vanishingly small close to optimal doping, but gets larger with reduced doping [156].

The same groups studied the isotope effect in another system of the Fe-based superconductors, namely in $Ba_{0.6}K_{0.4}Fe_2As_2$. In this system, however, both groups studied samples with the same $T_c \simeq 37\,K$. Again they observed a substantially different isotope effect: The same group finding the positive value in Sm1111 sees again a positive one with a similar value of $\alpha \simeq 0.37$ [87]. This is in severe antithesis to the negative value $\alpha \simeq -0.18$ seen by the other group [89]. This discrepancy in experimental results rises serious questions, which are partly clarified in the following.

Shortly after these works we also found a positive Fe isotope effect on T_c in $FeSe_{1-x}$ with and isotope exponent $\alpha = 0.81(15)$ much larger than the BCS value 0.5 [90]. For this study a series of samples with a

Chapter 5. Isotope effect

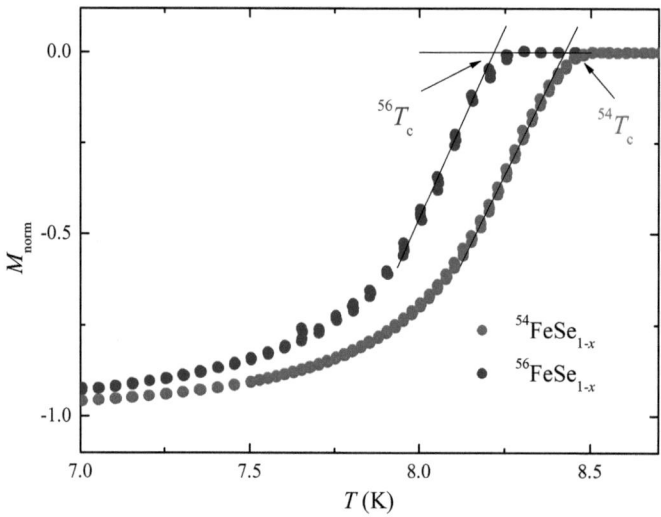

Figure 5.1: Normalized zero field cooled magnetization curves $M_{\text{norm}} = [M(T) - M_{\text{magn}}]/[M(2K) - M_{\text{magn}}]$ for one pair of ^{54}FeSe$_{1-x}$ and ^{56}FeSe$_{1-x}$ samples. The transition temperature T_c is determined as the intersection of the linearly extrapolated $M_{\text{norm}}(T)$ curve in the vicinity of T_c with the $M = 0$ line. After [90].

nominal composition of FeSe$_{0.98}$ were prepared by a solid state reaction described in Ref. [90]. The superconducting transition temperature of six ^{54}FeSe$_{1-x}$ and seven ^{56}FeSe$_{1-x}$ samples was investigated. Figure 5.1 shows a typical pair of magnetization curves for the two different Fe isotopes whereas the curve for the lighter ^{54}Fe is shifted almost parallel to higher temperatures, indicating that T_c of ^{54}FeSe$_{1-x}$ is higher than that of ^{56}FeSe$_{1-x}$. The superconducting transition temperatures for all the samples investigated are in two specific regions for the two different Fe isotopes in Fig. 5.2 with the corresponding mean values of $\overline{T_c}(^{54}\text{FeSe}_{1-x}) = 8.43(3)$ K and $\overline{T_c}(^{56}\text{FeSe}_{1-x}) = 8.21(4)$ K. This leads to the unusual large Fe isotope effect exponent of $\alpha = 0.81(15)$ mentioned above.

5.1. Isotope effect in the Fe-based superconductors

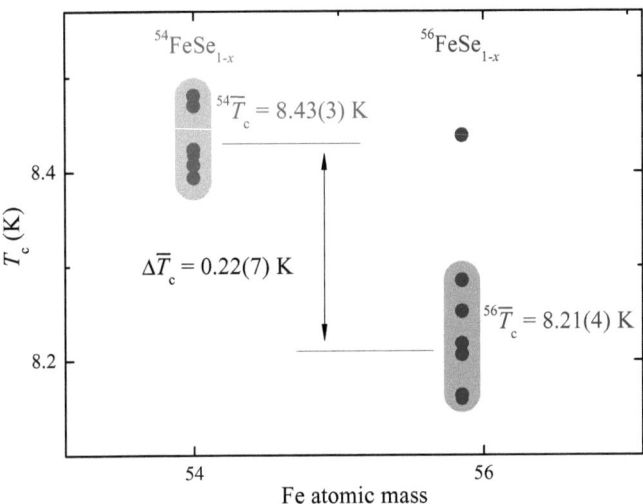

Figure 5.2: The superconducting transition temperature T_c as a function of Fe atomic mass for the ^{54}FeSe$_{1-x}$/^{56}FeSe$_{1-x}$ samples studied. The T_cs fall into the regions marked by the colored stripes with the corresponding mean values $^{54}\overline{T_c} = 8.43(3)$ K and $^{56}\overline{T_c} = 8.21(4)$ K. After [90].

Neutron powder diffraction on a pair of ^{54}Fe/^{56}Fe isotope substituted samples revealed that the lattice parameters are slightly different. Whereas the ^{54}FeSe$_{1-x}$ has a slightly larger $a(b)$-axis, the c-axis is smaller compared to ^{56}FeSe$_{1-x}$. This results in a shift of the chalcogenide height h above the Fe-plane. It has been shown empirically that only minor changes of the crystal structure, namely the anion height above the Fe-plane may change T_c substantially [54, 90]. An estimation of the change of T_c due to the isotope exchange was done with the help of pressure studies on FeSe$_{1-x}$. The decrease of the Se height, caused by a compression of the Fe$_4$Se pyramide, leads to an increase in T_c by $\delta T_c^h/(\delta h/h) \simeq 3.4\,\%/\text{K}$ [54, 68]. In addition, an increase in the Se(Te)-Fe-Se(Te) angle β in the Fe$_y$Se$_x$Te$_{1-x}$ family (in the superconducting regime of the phase diagram, $x \lesssim 0.5$) results in a decrease in T_c by $\delta T_c^\beta/(\delta \beta/\beta) \simeq 2.9\,\%/\text{K}$

55

Chapter 5. Isotope effect

[90]. From these observations a shift in $\delta T_c \approx 0.1\,\text{K}$ due to the structural effects is approximated. Thus $\approx 50\,\%$ of the increase of T_c in the lighter $^{54}\text{FeSe}_{1-x}$ compared to $^{56}\text{FeSe}_{1-x}$ are due to the change of the anion height above the Fe-plane [90]. Taking this additional "structural" isotope effect into account, one may write the total isotope exponent as a sum of two contributions:

$$\alpha_{\text{Fe}} = \alpha_{\text{Fe}}^{\text{int}} + \alpha_{\text{Fe}}^{\text{str}}. \qquad (5.3)$$

Here $\alpha_{\text{Fe}}^{\text{int}} = -(\delta T_c/T_c)/(\delta M/M)$ is the "intrinsic" component of the isotope effect exponent and $\alpha_{\text{Fe}}^{\text{str}} = -(\delta T_c/T_c)/(\delta h/h)$ is the contribution related to the change of the lattice parameters. This results in an "intrinsic" iron isotope exponent in $^{54/56}\text{FeSe}_{1-x}$ with a value of $\alpha_{\text{Fe}}^{\text{int}} \approx 0.4$ [90].

The use of the empirical T_c vs. h relation established in Ref. [54] allows to determine the structural related isotope effect. However, the absence of precise structural data complicates the analysis for the studies presented in Refs. [87, 88, 89]. But the intrinsic proportionality between h and the c-axis lattice parameter allows to determine the sign of the structural shift of T_c due to the isotope exchange [68, 99, 157]. Liu et al. [87] reported a zero effect on the lattice parameters within experimental accuracy for $\text{Ba}_{0.6}\text{K}_{0.4}\text{Fe}_2\text{As}_2$ and in $\text{SmFeAsO}_{0.85}\text{F}_{0.15}$. Thus $\alpha_{\text{Fe}}^{\text{str}} \sim 0$ for both studies. This leads to an "intrinsic" $\alpha_{\text{Fe}}^{\text{int}}$ in the direction of $0.35 - 0.4$, comparable to $\alpha_{\text{Fe}}^{\text{int}} \approx 0.4$ in FeSe_{1-x}, see Fig. 5.3. In contrary to Liu et al. [87], Shirage et al. [89] reported a negative isotope effect exponent α_{Fe} in the nominally identical compound $\text{Ba}_{0.6}\text{K}_{0.4}\text{Fe}_2\text{As}_2$. In this case, however, the c-axis lattice parameters differ from each other after the isotope exchange. Consequently, the negative effect stems from the sum of the "structural" and the "intrinsic" effect $(-0.18(\alpha_{\text{Fe}}) = 0.35(\alpha_{\text{Fe}}^{\text{int}}) - 0.53(\alpha_{\text{Fe}}^{\text{str}})$, see Eq. (5.3)). In the case of SmFeAsO_{1-x} [88], however, there are no isotope studies available with the same doping level. But by comparing the c-axis lattice parameters of the isotope exchanged samples, it is possible to estimate the sign of $\alpha_{\text{Fe}}^{\text{str}}$. Based on the other Fe isotope studies [87, 90, 91], also in this case an "intrinsic" Fe isotope effect exponent of $\alpha_{\text{Fe}}^{\text{int}} \sim 0.35 - 0.4$ can be

5.1. Isotope effect in the Fe-based superconductors

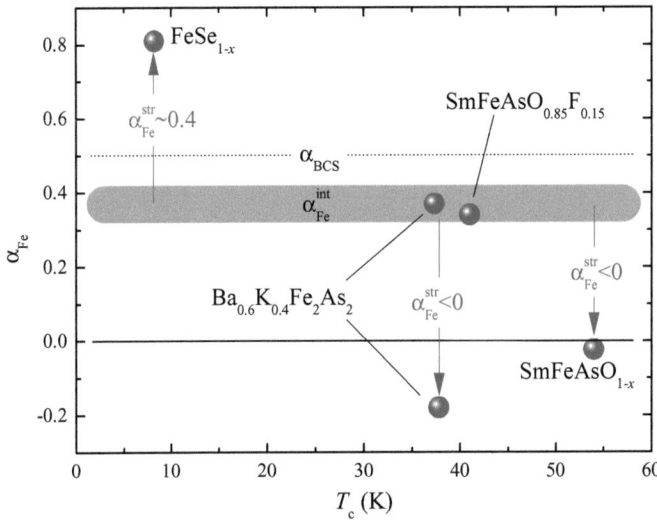

Figure 5.3: Fe isotope effect exponent α_{Fe} as a function of the superconducting transition temperature T_c for the samples considered in the present study: $FeSe_{1-x}$ [90], $Ba_{0.6}K_{0.4}Fe_2As_2$ and $SmFeAsO_{0.85}F_{0.15}$ [87], $Ba_{0.6}K_{0.4}Fe_2As_2$ [88], and $SmFeAsO_{1-x}$ [89]. Arrows indicate the direction of the shift from the "intrinsic" Fe isotope effect exponent $\alpha_{Fe}^{int} \sim 0.35 - 0.4$ caused by the structural effects. $\alpha_{BCS} \equiv 0.5$ is the BCS value for electron-phonon mediated superconductivity. After [91].

proposed, see Fig. 5.3 [91].
Recently Bussmann-Holder *et al.* [158] investigated a multiple-gap scenario of superconductivity in the Fe-based HTC superconductors to find possible sources of the isotope effect on T_c. Electron-phonon mediated superconductivity within the dominant gap channel predicts a T_c independent isotope effect with a value of α slightly smaller than 0.5. This is in agreement with the observed $\alpha_{Fe}^{int} \sim 0.35 - 0.4$ for superconductors belonging to different families of the Fe-based HTC superconductors after the correction for the structural effect and with T_c's ranging from 8 K to 54 K (see Fig. 5.3).

57

Chapter 5. Isotope effect

5.2 Related publications to Chapter 5

5.2.1 Paper I: Iron isotope effect on the superconducting transition temperature and the crystal structure of FeSe$_{1-x}$

This work is published in:

R. Khasanov, M. Bendele, K. Conder, H. Keller, E. Pomjakushina, and V. Pomjakushin, *Iron isotope effect on the superconducting transition temperature and the crystal structure of FeSe$_{1-x}$*, New J. Phys., **12**, 073024 (2010).

Abstract:

The Fe isotope effect (Fe-IE) on the transition temperature T_c and the crystal structure was studied in the Fe chalcogenide superconductor FeSe$_{1-x}$ by means of magnetization and neutron powder diffraction (NPD). The substitution of natural Fe (containing $\simeq$ 92% of ^{56}Fe) by its lighter ^{54}Fe isotope leads to a shift in T_c of 0.22(5) K corresponding to an Fe-IE exponent of $\alpha_{Fe} = 0.81(15)$. Simultaneously, a small structural change with isotope substitution is observed by NPD, which may contribute to the total Fe isotope shift of T_c.

URL: http://iopscience.iop.org/1367-2630/12/7/073024
DOI: 10.1088/1367-2630/12/7/073024
PACS: 74.62.Dh, 74.25.Ha, 74.62.Bf, 74.72.-h, 75.60.Ej
©IOP Publishing & Deutsche Physikalische Gesellschaft. This article is available under a CC BY-NC-SA licence from the NJP website

5.2.2 Paper II: Intrinsic and structural isotope effects in Fe-based superconductors

This work is published in:

R. Khasanov, M. Bendele, A. Bussmann-Holder, and H. Keller, *Intrinsic and structural isotope effects in Fe-based superconductors*, Phys. Rev. B **82**, 212505 (2010).

Abstract:

The currently available results of the isotope effect on the superconducting transition temperature T_c in Fe-based high-temperature superconductors (HTSs) are highly controversial. The values of the Fe isotope effect exponent α_{Fe} for various families of Fe-based HTS were found to be as well positive, as negative, or even be exceedingly larger than the BCS value $\alpha_{BCS} \equiv 0.5$. Here we emphasize that the Fe isotope substitution causes small structural modifications which, in turn, affect T_c. Upon correcting the isotope effect exponent for these structural effects, an almost unique value of $\alpha \sim 0.35 - 0.4$ is observed for at least three different families of Fe-based HTS.

URL: http://link.aps.org/doi/10.1103/PhysRevB.82.212505
DOI: 10.1103/PhysRevB.82.212505
PACS: 74.70.Xa, 74.62.Bf, 74.25.Kc
Copyright 2010 by The American Physical Society.

6 Conclusion and Outlook

The investigations of the FeCh system by means of μSR, magnetization, and neutron diffraction measurements presented in this book provide a detailed insight into the superconducting and magnetic properties of the system.

The hydrostatic pressure effect on the superconducting and magnetic properties of the binary FeSe$_{1-x}$ was studied. The system exhibits one of the highest known pressure effects on the superconducting transition temperature T_c and an unusual pressure dependence with a local maximum at $p \simeq 0.8$ GPa and a local minimum at a slightly higher pressure of $p \simeq 1.2$ GPa. The system is bulk superconducting within the pressure range investigated. The μSR experiments reveal the appearance of magnetism that coexist with superconductivity above $p \simeq 0.8$ GPa whereat the Néel temperature $T_N > T_c$. In a narrow pressure range where the local minimum in $T_c(p)$ is observed, magnetism competes with superconductivity in the sense that T_c is decreasing, and the magnetic volume fraction and the internal magnetic field decrease below T_c. However, at pressures above the local minimum of $T_c(p)$, superconductivity and magnetism coexist on short length scales in the entire sample volume with superconducting and magnetic volume fractions reaching both 100%. In addition both T_c and T_N as well as the magnetic order parameter increase with increasing pressure. This extraordinary behavior needs further investigation and provides a new challenge for theories of superconductivity.

The substitution of Se by Te in FeSe$_{1-x}$ induces a chemical pressure effect. The resulting phase diagram of Fe$_y$Se$_x$Te$_{1-x}$ is very similar to that of other Fe-based superconductors. First T_c increases until at an intermediate range of substitution superconductivity and incommensurate magnetism coexist. Finally, the fully substituted FeTe is a commensurate antiferromagnet. Furthermore, the superconducting and magnetic properties depend strongly on the Fe content of the system in the coexistence region of superconductivity and magnetism. In the low Fe content region bulk superconductivity and incommensurate magnetism coexist. With increasing y the magnetic order becomes more correlated over a longer range and superconductivity vanishes. It will be important to investigate the role of excess Fe and accompanying Fe vacancies and how the local atomic structure and the Fermi-surface are affected.

The μSR studies of superconducting FeSe$_{0.5}$Te$_{0.5}$ indicate multi-gap superconductivity as observed for most Fe-based superconductors. Comparing the Fe-based high temperature (HTC) superconductors with other classes suggests that the two gap scenario is generic for high temperature superconductors while the order parameter is different in every class: In MgB$_2$ and most probably in the Fe-based superconductors it is $s+s$-wave like, whereas in the latter case the situation is not fully resolved yet whether the fully gaped s-wave states have a different phase or not ($s_{\pm}$- or s_{++}-wave). It is even possible, that the Fe-based superconductors have a nodal $s_{\pm}$ gap structure. In the cuprate HTC superconductors, on the other hand, nodes are present in the gap structure that is $d+s$-wave like. However, even if the symmetry of the gap-structure is known, one cannot conclude on the nature of the pairing mechanism in the Fe-based superconductors.

Isotope effect studies show a significant difference in T_c after the iron-isotope substitution (^{54}Fe/^{56}Fe) in all families of Fe-based superconductors. The isotope effect exponents reach values as large as $\alpha_{Fe} = 0.81(15)$ in FeSe$_{1-x}$. However, a minor change in the lattice parameters was observed in the isotope exchanged samples. It turned out that T_c is extremely sensitive to any changes in the lattice parameters, in particular the anion (Se) height above the Fe-plane. Taking this structural effect into account, α_{Fe} reduces to an intrinsic isotope exponent $\alpha_{Fe}^{int} \simeq 0.4$. Careful comparison of the lattice structure in the other families of Fe-based superconductors, resulted in intrinsic isotope exponents of comparable values: $\alpha_{Fe}^{int} \simeq 0.35-0.4$. These findings clearly show a conventional isotope effect on T_c and highlight the role of the lattice in the pairing mechanism in this new material class. However, only four experimental studies are available up to now. It will be interesting to gain more detailed information on the local atomic structure and binding valence in order to get more insight on the origin of the structural isotope effect. Furthermore, the impact of the anions (As/P and Se/Te) on the pairing must be tested with isotope exchange experiments.

In conclusion, the FeCh system investigated in the framework of the present book shows a substantial sensitivity of the superconducting and magnetic properties for small variations of various physical parameters of the system. Application of hydrostatic pressure drives T_c up to 36 K and commensurate magnetism occurs in the system, coexisting with superconductivity. Chemical pressure, induced by substitution of Se with the isovalent Te, leads to similar effects, whereas the FeTe parent compound is not superconducting. Even the isotope substitution of ^{56}Fe by ^{54}Fe gives rise to slightly different lattice parameters that substantially contribute to the isotope exponent. Combining all these observations made for the FeCh system leads to the conclusion that the lattice and its parameters play a major, if not the most important role in determining whether the system is superconducting, magnetically ordered, or even in a state where both phenomena are coexisting on a microscopic scale.

More than three year after the discovery of the Fe-based superconductors, the origin of superconductivity in this novel class of superconductors remains a challenging field of modern condensed matter physics research. At the end it remains to be seen whether superconductivity in these classes of high-temperature superconductors has the same or a similar origin.

Bibliography

[1] H. Kamerlingh Onnes, *The liquefaction of helium*, Proc. K. Akad. Amsterdam **11**, 168 (1908).

[2] H. Kamerlingh Onnes, *Further experiments with liquid helium D - On the change of the electrical resistance of pure metals at very low temperatures, etc. V The disappearance of the resistance of mercury*, Proc. K. Akad. Amsterdam **14**, 113 (1911).

[3] W. Meissner and R. Ochsenfeld, *Short initial announcements*, Naturwissenschaften **21**, 787 (1933).

[4] F. London and H. London, *The electromagnetic equations of the supraconductor*, Proc. R. Soc. A **149**, 71 (1935).

[5] F. London and H. London, *Supra conduction and diamagnetism*, Physica (Amsterdam) **2**, 341 (1935).

[6] V. L. Ginzburg and L. D. Landau, *On the theory of superconductivity*, Zh. Eksp. Theor. Fiz. **20**, 1064 (1950).

[7] A. Abrikosov, *On the magnetic properties of superconductors of the second group*, Sov. Phys. JETP **5**, 1174 (1957).

[8] E. Maxwell, *Isotope effect in the superconductivity of mercury*, Phys. Rev. **78**, 477 (1950).

[9] C. A. Reynolds, B. Serin, W. H. Wright, and L. B. Nesbitt, *Superconductivity of isotopes of mercury*, Phys. Rev. **78**, 487 (1950).

[10] J. Bardeen, L. N. Cooper, and J. R. Schrieffer, *Microscopic theory of superconductivity*, Phys. Rev. **106**, 162 (1957).

Bibliography

[11] J. Bardeen, L. N. Cooper, and J. R. Schrieffer, *Theory of superconductivity*, Phys. Rev. **108**, 1175 (1957).

[12] L. P. Gorkov, *Microscopic deruvation of the Ginzburg-Landay equations int the theory of superconductivity*, Soviet Physics JETP-USSR **9**, 1364 (1959).

[13] J. P. Carbotte, *Properties of boson-exchange superconductors*, Rev. Mod. Phys. **62**, 1027 (1990).

[14] J. G. Bednorz and K. A. Müller, *Possible high-T_c superconductivity in the Ba-La-Cu-O system*, Z. Phys. B - Condens. Matter **64**, 189 (1986).

[15] A. Schilling, M. Cantoni, J. D. Guo, and H. R. Ott, *Superconductivity above 130 K in the Hg-Ba-Ca-Cu-O system*, Nature (London) **363**, 56 (1993).

[16] L. Gao, Y. Y. Xue, F. Chen, Q. Xiong, R. L. Meng, D. Ramirez, C. W. Chu, J. H. Eggert, and H. K. Mao, *Superconductivity up to 164 K in $HgBa_2Ca_{m-1}Cu_mO_{2m+2+\delta}$ (m=1, 2, and 3) under quasihydrostatic pressures*, Phys. Rev. B **50**, 4260 (1994).

[17] A. Bussmann-Holder and H. Keller, *Polaron formation as origin of unconventional isotope effects in cuprate superconductors*, Eur. Phys. J. B **44**, 487 (2005).

[18] A. Bussmann-Holder, H. Keller, A. R. Bishop, A. Simon, R. Micnas, and K. A. Müller, *Unconventional isotope effects as evidence for polaron formation in cuprates*, EPL **72**, 423 (2005).

[19] H. Keller, A. Bussmann-Holder, and K. A. Müller, *Jahn-Teller physics and high-T_c superconductivity*, Materials Today **11**, 38 (2008).

[20] K. A. Müller, *On the superconductivity in hole doped cuprates*, J. Phys.: Condens. Matter **19**, 251002 (2007).

[21] J. Nagamatsu, N. Nakagawa, T. Muranaka, Y. Zenitani, and J. Akimitsu, *Superconductivity at 39 K in magnesium diboride*, Nature (London) **410**, 63 (2001).

[22] A. Y. Liu, I. I. Mazin, and J. Kortus, *Beyond Eliashberg superconductivity in MgB_2: Anharmonicity, two-phonon scattering, and multiple gaps*, Phys. Rev. Lett. **87**, 087005 (2001).

[23] Y. Kamihara, T. Watanabe, M. Hirano, and H. Hosono, *Iron-based layered superconductor $La[O_{1-x}F_x]FeAs$ (x=0.05-0.12) with $T_c=26$ K*, J. Am. Chem. Soc. **130**, 3296 (2008).

[24] P. J. Hirschfeld, M. M. Korshunov, and I. I. Mazin, *Gap symmetry and structure of Fe-based superconductors*, Reports on Progress in Physics **74**, 124508 (2011).

[25] A. Subedi, L. Zhang, D. J. Singh, and M. H. Du, *Density functional study of FeS, FeSe, and FeTe: Electronic structure, magnetism, phonons, and superconductivity*, Phys. Rev. B **78**, 134514 (2008).

[26] M. Tinkham, *Introduction to superconductivity*, McGraw-Hill, New York, 1996.

[27] W. Buckel and R. Kleiner, *Supraleitung: Grundlagen und Anwendungen*, Wiley-VCH, Wernheim, 2004.

[28] T. Schneider, *The physics of superconductors*, edited by K. Bennemann and J. B. Ketterson, Springer, Berlin, 2004.

[29] L. D. Landau, *The theory of phase transitions*, Nature (London) **138**, 840 (1936).

[30] L. N. Cooper, *Bound electron pairs in a degenerate Fermi gas*, Phys. Rev. **104**, 1189 (1956).

[31] A. Schenck, *Muon Spin Rotation Spectroscopy: Principles and Applications in Solid State Physics*, Adam Hilger, Bristol, 1985.

[32] A. Yaouanc and P. D. de Réotier, *Muon Spin Rotation, Relaxation, and Resonance*, Oxford Science Publications, 2011.

[33] R. Kubo and T. Toyabe, *A stochastic model for low field resonance and relaxation*, in A stochastic model for low field resonance and relaxation, pages 810–823, North-Holland, Amsterdam, 1967.

[34] J. Major, J. Mundy, M. Schmolz, A. Seeger, K. Döring, K. Fürderer, M. Gladisch, D. Herlach, and G. Majer, *Zero-field muon spin rotation in monocrystalline chromium*, Hyperfine Interactions **31**, 259 (1986).

[35] A. Amato, *Heavy-fermion systems studied by µSR technique*, Rev. Mod. Phys. **69**, 1119 (1997).

[36] R. S. Hayano, Y. J. Uemura, J. Imazato, N. Nishida, T. Yamazaki, and R. Kubo, *Zero-and low-field spin relaxation studied by positive muons*, Phys. Rev. B **20**, 850 (1979).

[37] J. E. Sonier, J. H. Brewer, and R. F. Kiefl, *µSR studies of the vortex state in type-II superconductors*, Rev. Mod. Phys. **72**, 769 (2000).

[38] A. Maisuradze, R. Khasanov, A. Shengelaya, and H. Keller, *Comparison of different methods for analyzing SR line shapes in the vortex state of type-II superconductors*, J. Phys.: Condens. Matter **21**, 075701 (2009).

[39] E. H. Brandt, *Flux distribution and penetration depth measured by muon spin rotation in high-Tc superconductors*, Phys. Rev. B **37**, 2349 (1988).

[40] Z. Hao, J. R. Clem, M. W. McElfresh, L. Civale, A. P. Malozemoff, and F. Holtzberg, *Model for the reversible magnetization of high-κ type-II superconductors: Application to high-T_c superconductors*, Phys. Rev. B **43**, 2844 (1991).

[41] A. Yaouanc, P. Dalmas de Réotier, and E. H. Brandt, *Effect of the vortex core on the magnetic field in hard superconductors*, Phys. Rev. B **55**, 11107 (1997).

[42] E. H. Brandt, *Magnetic field density of perfect and imperfect flux line lattices in type II superconductors. I. Application of periodic solutions*, Journal of Low Temperature Physics **73**, 355 (1988).

[43] Y. Kamihara, H. Hiramatsu, M. Hirano, R. Kawamura, H. Yanagi, T. Kamiya, and H. Hosono, *Iron-based layered superconductor: LaOFeP*, J. Am. Chem. Soc. **128**, 10012 (2006).

[44] B. S. Chandrasekhar and J. K. Hulm, *The electrical resistivity and superconductivity of some uranium alloys and compounds*, J. Phys. Chem. Solids **7**, 259 (1958).

[45] I. Shirotani, Y. Shimaya, K. Kihou, C. Sekine, N. Takeda, M. Ishikawa, and T. Yagi, *Superconductivity of new filled skutterudite YFe_4P_{12} prepared at high pressure*, J. Phys.: Condens. Matter **15**, 2201 (2003).

[46] G. P. Meisner, *Superconductivity and magnetic order in ternary rare-earth transition metal phosphides*, Physica B & C **108**, 763 (1981).

[47] H. Takahashi, K. Igawa, K. Arii, Y. Kamihara, M. Hirano, and H. Hosono, *Superconductivity at 43 K in an iron-based layered compound $LaO_{1-x}F_xFeAs$*, Nature (London) **453**, 376 (2008).

[48] V. Johnson and W. Jeitschko, *ZrCiSiAs - Filled PbFCl-type*, J. Solid State Chem. **11**, 161 (1974).

[49] X. H. Chen, T. Wu, G. Wu, R. H. Liu, H. Chen, and D. F. Fang, *Superconductivity at 43 K in $SmFeAsO_{1-x}F_x$*, Nature (London) **453**, 761 (2008).

[50] Z.-A. Ren, J. Yang, W. Lu, W. Yi, X.-L. Shen, Z.-C. Li, G.-C. Che, X.-L. Dong, L.-L. Sun, F. Zhou, and Z.-X. Zhao, *Superconductivity in the iron-based F-doped layered quaternary compound $Nd[O_{1-x}F_x]FeAs$*, EPL **82**, 57002 (2008).

Bibliography

[51] G. F. Chen, Z. Li, D. Wu, G. Li, W. Z. Hu, J. Dong, P. Zheng, J. L. Luo, and N. L. Wang, *Superconductivity at 41 K and its competition with spin-density-wave instability in layered $CeO_{1-x}F_xFeAs$*, Phys. Rev. Lett. **100**, 247002 (2008).

[52] Z.-A. Ren, G.-C. Che, X.-L. Dong, J. Yang, W. Lu, W. Yi, X.-L. Shen, Z.-C. Li, L.-L. Sun, F. Zhou, and Z.-X. Zhao, *Superconductivity and phase diagram in iron-based arsenic-oxides $ReFeAsO_{1-\delta}$ (Re = rare-earth metal) without fluorine doping*, EPL **83**, 17002 (2008).

[53] J. Karpinski, N. D. Zhigadlo, S. Katrych, Z. Bukowski, P. Moll, S. Weyeneth, H. Keller, R. Puzniak, M. Tortello, D. Daghero, R. Gonnelli, I. Maggio-Aprile, Y. Fasano, O. Fischer, K. Rogacki, and B. Batlogg, *Single crystals of $LnFeAsO_{1-x}F_x$ (Ln = La, Pr, Nd, Sm, Gd) and $Ba_{1-x}Rb_xFe_2As_2$: Growth, structure and superconducting properties*, Physica C **469**, 370 (2009).

[54] Y. Mizuguchi, Y. Hara, K. Deguchi, S. Tsuda, T. Yamaguchi, K. Takeda, H. Kotegawa, H. Tou, and Y. Takano, *Anion height dependence of T_c for the Fe-based superconductor*, Supercond. Sci. Technol. **23**, 054013 (2010).

[55] M. Rotter, M. Tegel, and D. Johrendt, *Superconductivity at 38 K in the iron arsenide $(Ba_{1-x}K_x)Fe_2As_2$*, Phys. Rev. Lett. **101**, 107006 (2008).

[56] K. Sasmal, B. Lv, B. Lorenz, A. M. Guloy, F. Chen, Y.-Y. Xue, and C.-W. Chu, *Superconducting Fe-based compounds $(A_{1-x}Sr_x)Fe_2As_2$ with A = K and Cs with transition temperatures up to 37 K*, Phys. Rev. Lett. **101**, 107007 (2008).

[57] G. Wu, H. Chen, T. Wu, Y. L. Xie, Y. J. Yan, R. H. Liu, X. F. Wang, J. J. Ying, and X. H. Chen, *Different resistivity response to spin-density wave and superconductivity at 20 K in $Ca_{1-x}Na_xFe_2As_2$*, J. Phys.: Condens. Matter **20**, 422201 (2008).

[58] H. S. Jeevan, Z. Hossain, D. Kasinathan, H. Rosner, C. Geibel, and P. Gegenwart, *High-temperature superconductivity in $Eu_{0.5}K_{0.5}Fe_2As_2$*, Phys. Rev. B **78**, 092406 (2008).

[59] Y. Qi, Z. Gao, L. Wang, D. Wang, X. Zhang, and Y. Ma, *Superconductivity at 34.7 K in the iron arsenide $Eu_{0.7}Na_{0.3}Fe_2As_2$*, New J. Phys. **10**, 123003 (2008).

[60] X. C. Wang, Q. Q. Liu, Y. X. Lv, W. B. Gao, L. X. Yang, R. C. Yu, F. Y. Li, and C. Q. Jin, *The superconductivity at 18 K in LiFeAs system*, Solid State Comm. **148**, 538 (2008).

[61] D. R. Parker, M. J. Pitcher, P. J. Baker, I. Franke, T. Lancaster, S. J. Blundell, and S. J. Clarke, *Structure, antiferromagnetism and superconductivity of the layered iron arsenide NaFeAs*, Chem. Comm. **16**, 2189 (2009).

[62] H. Ogino, Y. Matsumura, Y. Katsura, K. Ushiyama, S. Horii, K. Kishio, and J. Shimoyama, *Superconductivity at 17 K in $(Fe_2P_2)(Sr_4Sc_2O_6)$: A new superconducting layered pnictide oxide with a thick perovskite oxide layer*, Supercond. Sci. Technol. **22**, 075008 (2009).

[63] X. Zhu, F. Han, G. Mu, B. Zeng, P. Cheng, B. Shen, and H.-H. Wen, *$Sr_3Sc_2Fe_2As_2O_5$ as a possible parent compound for FeAs-based superconductors*, Phys. Rev. B **79**, 024516 (2009).

[64] F.-C. Hsu, J.-Y. Luo, K.-W. Yeh, T.-K. Chen, T.-W. Huang, P. M. Wu, Y.-C. Lee, Y.-L. Huang, Y.-Y. Chu, D.-C. Yan, and M.-K. Wu, *Superconductivity in the PbO-type structure α-FeSe*, Proc. Nat. Acad. Sci. USA **105**, 14262 (2008).

[65] K.-W. Yeh, T.-W. Huang, Y. l. Huang, T.-K. Chen, F.-C. Hsu, P. M. Wu, Y.-C. Lee, Y.-Y. Chu, C.-L. Chen, J.-Y. Luo, D.-C. Yan, and M.-K. Wu, *Tellurium substitution effect on superconductivity of the α-phase iron selenide*, EPL **84**, 37002 (2008).

[66] B. C. Sales, A. S. Sefat, M. A. McGuire, R. Y. Jin, D. Mandrus, and Y. Mozharivskyj, *Bulk superconductivity at 14 K in single crystals of $Fe_{1+y}Te_xSe_{1-x}$*, Phys. Rev. B **79**, 094521 (2009).

[67] Y. Mizuguchi, F. Tomioka, S. Tsuda, T. Yamaguchi, and Y. Takano, *Substitution Effects on FeSe Superconductor*, J. Phys. Soc. Jap. **78**, 074712 (2009).

[68] S. Margadonna, Y. Takabayashi, Y. Ohishi, Y. Mizuguchi, Y. Takano, T. Kagayama, T. Nakagawa, M. Takata, and K. Prassides, *Pressure evolution of the low-temperature crystal structure and bonding of the superconductor FeSe ($T_c = 37\,K$)*, Phys. Rev. B **80**, 064506 (2009).

[69] J. Guo, S. Jin, G. Wang, S. Wang, K. Zhu, T. Zhou, M. He, and X. Chen, *Superconductivity in the iron selenide $K_xFe_2Se_2$ ($0 \leq x \leq 1.0$)*, Phys. Rev. B **82**, 180520 (2010).

[70] A. Krzton-Maziopa, Z. Shermadini, E. Pomjakushina, V. Pomjakushin, M. Bendele, A. Amato, R. Khasanov, H. Luetkens, and K. Conder, *Synthesis and crystal growth of $Cs_{0.8}(FeSe_{0.98})_2$: a new iron-based superconductor with $T_c = 27\,K$*, Journal of Physics: Condensed Matter **23**, 052203 (2011).

[71] A. F. Wang, J. J. Ying, Y. J. Yan, R. H. Liu, X. G. Luo, Z. Y. Li, X. F. Wang, M. Zhang, G. J. Ye, P. Cheng, Z. J. Xiang, and X. H. Chen, *Superconductivity at 32 K in single-crystalline $Rb_xFe_{2-y}Se_2$*, Phys. Rev. B **83**, 060512 (2011).

[72] S. Weyeneth, *Anisotropic properties and critical behavior of high-temperature superconductors*, Ph.D. thesis, University of Zurich, 2009.

[73] L. Boeri, O. V. Dolgov, and A. A. Golubov, *Is $LaFeAsO_{1-x}F_x$ an Electron-Phonon Superconductor?*, Phys. Rev. Lett. **101**, 026403 (2008).

Bibliography

[74] D. V. Evtushinsky, D. S. Inosov, V. B. Zabolotnyy, M. S. Viazovska, R. Khasanov, A. Amato, H.-H. Klauss, H. Luetkens, C. Niedermayer, G. L. Sun, V. Hinkov, C. T. Lin, A. Varykhalov, A. Koitzsch, M. Knupfer, B. Büchner, A. A. Kordyuk, and S. V. Borisenko, *Momentum-resolved superconducting gap in the bulk of $Ba_{1-x}K_xFe_2As_2$ from combined ARPES and μSR measurements*, New Journal of Physics **11**, 055069 (2009).

[75] K. Kuroki, H. Usui, S. Onari, R. Arita, and H. Aoki, *Pnictogen height as a possible switch between high-Tc nodeless and low-Tc nodal pairings in the iron-based superconductors*, Phys. Rev. B **79**, 224511 (2009).

[76] K. Seo, B. A. Bernevig, and J. Hu, *Pairing Symmetry in a Two-Orbital Exchange Coupling Model of Oxypnictides*, Phys. Rev. Lett. **101**, 206404 (2008).

[77] S. Onari, H. Kontani, and M. Sato, *Structure of neutron-scattering peaks in both s_{++}-wave and $s_{\pm}$-wave states of an iron pnictide superconductor*, Phys. Rev. B **81**, 060504 (2010).

[78] I. I. Mazin, D. J. Singh, M. D. Johannes, and M. H. Du, *Unconventional superconductivity with a sign reversal in the order parameter of $LaFeAsO_{1-x}F_x$*, Phys. Rev. Lett. **101**, 057003 (2008).

[79] A. V. Chubukov, *Renormalization group analysis of competing orders and the pairing symmetry in Fe-based superconductors*, Physica C **469**, 640 (2009).

[80] A. V. Chubukov, M. G. Vavilov, and A. B. Vorontsov, *Momentum dependence and nodes of the superconducting gap in the iron pnictides*, Phys. Rev. B **80**, 140515 (2009).

[81] A. B. Vorontsov, M. G. Vavilov, and A. V. Chubukov, *Superconductivity and spin-density waves in multiband metals*, Phys. Rev. B **81**, 174538 (2010).

[82] R. M. Fernandes and J. Schmalian, *Competing order and nature of the pairing state in the iron pnictides*, Phys. Rev. B **82**, 014521 (2010).

[83] A. B. Vorontsov, M. G. Vavilov, and A. V. Chubukov, *Interplay between magnetism and superconductivity in the iron pnictides*, Phys. Rev. B **79**, 060508 (2009).

[84] T. Yildirim, *Origin of the 150-K Anomaly in LaFeAsO: Competing Antiferromagnetic Interactions, Frustration, and a Structural Phase Transition*, Phys. Rev. Lett. **101**, 057010 (2008).

[85] T. Yildirim, *Frustrated magnetic interactions, giant magnetoelastic coupling, and magnetic phonons in iron-pnictides*, Physica C **469**, 425 (2009).

[86] F. Yndurain and J. M. Soler, *Anomalous electron-phonon interaction in doped LaFeAsO: First-principles calculations*, Phys. Rev. B **79**, 134506 (2009).

[87] H. Liu, T. Wu, H. Chen, X. F. Wang, Y. L.Xie, J. J. Ying, Y. J. Yan, Q. J. Li, B. C. Shi, W. S. Chu, Z. Y. Wu, and X. H. Chen, *A large iron isotope effect in $SmFeAsO_{1-x}F_x$ and $Ba_{1-x}K_xFe_2As_2$*, Nature (London) **459**, 64 (2008).

[88] P. M. Shirage, K. Miyazawa, K. Kihou, H. Kito, Y. Yoshida, Y. Tanaka, H. Eisaki, and A. Iyo, *Absence of an Appreciale Iron Isotope Effect on the Transition Temperature of the Optimally Doped $SmFeAsO_{1-y}$ Superconductor*, Phys. Rev. Lett. **105**, 037004 (2010).

[89] P. M. Shirage, K. Kihou, K. Miyazawa, C.-H. Lee, H. Kito, H. Eisaki, T. Yanagisawa, Y. Tanaka, and A. Iyo, *Inverse Iron Isotope Effect on the Transition Temperature of the $(Ba,K)Fe_2As_2$ Superconductor*, Phys. Rev. Lett. **103**, 257003 (2009).

[90] R. Khasanov, M. Bendele, K. Conder, H. Keller, E. Pomjakushina, and V. Pomjakushin, *Iron isotope effect on the superconducting*

transition temperature and the crystal structure of $FeSe_{1-x}$, New J. Phys. **12**, 073024 (2010).

[91] R. Khasanov, M. Bendele, A. Bussmann-Holder, and H. Keller, *Intrinsic and structural isotope effects in Fe-based superconductors*, Phys. Rev. B **82**, 212505 (2010).

[92] K. Terashima, Y. Sekiba, J. H. Bowen, K. Nakayama, T. Kawahara, T. Sato, P. Richard, Y.-M. Xu, L. J. Li, G. H. Cao, Z.-A. Xu, H. Ding, and T. Takahashi, *Fermi surface nesting induced strong pairing in iron-based superconductors*, Proc. Nat. Acad. Sci. USA **106**, 7330 (2009).

[93] D. C. Johnston, *The puzzle of high temperature superconductivity in layered iron pnictides and chalcogenides*, Advances In Physics **59**, 803 (2010).

[94] M. D. Lumsden and A. D. Christianson, *Magnetism in Fe-based superconductors*, J. Phys.: Condens. Matter **22**, 203203 (2010).

[95] S.-H. Lee, G. Xu, W. Ku, J. S. Wen, C. C. Lee, N. Katayama, Z. J. Xu, S. Ji, Z. W. Lin, G. D. Gu, H.-B. Yang, P. D. Johnson, Z.-H. Pan, T. Valla, M. Fujita, T. J. Sato, S. Chang, K. Yamada, and J. M. Tranquada, *Coupling of spin and orbital excitations in the iron-based superconductor* $FeSe_{0.5}Te_{0.5}$, Phys. Rev. B **81**, 220502 (2010).

[96] P. Babkevich, M. Bendele, A. T. Boothroyd, K. Conder, S. N. Gvasaliya, R. Khasanov, E. Pomjakushina, and B. Roessli, *Magnetic excitations of* $Fe_{1+y}Se_xTe_{1-x}$ *in magnetic and superconductive phases*, J. Phys.: Condens. Matter **22**, 142202 (2010).

[97] J. Zhang, R. Sknepnek, R. M. Fernandes, and J. Schmalian, *Orbital coupling and superconductivity in the iron pnictides*, Phys. Rev. B **79**, 220502 (2009).

[98] E. Pomjakushina, K. Conder, V. Pomjakushin, M. Bendele, and R. Khasanov, *Synthesis, crystal structure, and chemical stability of the superconductor* $FeSe_{1-x}$, Phys. Rev. B **80**, 024517 (2009).

Bibliography

[99] T. M. McQueen, Q. Huang, V. Ksenofontov, C. Felser, Q. Xu, H. Zandbergen, Y. S. Hor, J. Allred, A. J. Williams, D. Qu, J. Checkelsky, N. P. Ong, and R. J. Cava, *Extreme sensitivity of superconductivity to stoichiometry in $Fe_{1+\delta}Se$*, Phys. Rev. B **79**, 014522 (2009).

[100] S. Margadonna, Y. Takabayashi, M. T. McDonald, K. Kasperkiewicz, Y. Mizuguchi, Y. Takano, A. N. Fitch, E. Suard, and K. Prassides, *Crystal structure of the new $FeSe_{1-x}$ superconductor*, Chem. Commun. , 5607 (2008).

[101] S. Medvedev, T. M. McQueen, I. A. Troyan, T. Palasyuk, M. I. Eremets, R. J. Cava, S. Naghavi, F. Casper, V. Ksenofontov, G. Wortmann, and C. Felser, *Electronic and magnetic phase diagram of β-$Fe_{1.01}Se$ with superconductivity at 36.7 K under pressure*, Nature Materials **8**, 630 (2009).

[102] M. Bendele, A. Amato, K. Conder, M. Elender, H. Keller, H.-H. Klauss, H. Luetkens, E. Pomjakushina, A. Raselli, and R. Khasanov, *Pressure Induced Static Magnetic Order in Superconducting $FeSe_{1-x}$*, Phys. Rev. Lett. **104**, 087003 (2010).

[103] K. Miyoshi, Y. Takaichi, E. Mutou, K. Fujiwara, and J. Takeuchi, *Anomalous Pressure Dependence of the Superconducting Transition Temperature in $FeSe_{1-x}$ Studied by DC Magnetic Measurements*, J. Phys. Soc. Jap. **78**, 093703 (2009).

[104] S. Masaki, H. Kotegawa, Y. Hara, H. Tou, K. Murata, Y. Mizuguchi, and Y. Takano, *Precise Pressure Dependence of the Superconducting Transition Temperature of FeSe: Resistivity and ^{77}Se-NMR Study*, J. Phys. Soc. Jap. **78**, 063704 (2009).

[105] M. Bendele, A. Ichsanow, Y. Pashkevich, L. Keller, T. Strässle, A. Gusev, E. Pomjakushina, K. Conder, R. Khasanov, and H. Keller, *Coexistence of superconductivity and magnetism in $FeSe_{1-x}$ under pressure*, Phys. Rev. B **85**, 064517 (2012).

Bibliography

[106] A. Ichsanow, *Magnetic and Superconducting properties of $FeSe_{1-x}$, dependency on sample preparation*, Master thesis, University of Zurich, 2010.

[107] A. Bussmann-Holder, R. Micnas, and A. Bishop, *Enhancements of the superconducting transition temperature within the two-band model*, Eur. Phys. J. B **37**, 345 (2004).

[108] V. G. Kogan, C. Martin, and R. Prozorov, *Superfluid density and specific heat within a self-consistent scheme for a two-band superconductor*, Phys. Rev. B **80**, 014507 (2009).

[109] A. Bussmann-Holder, *Comment on: Superfluid density and specific heat within a self-consistent scheme for a two-band superconductor*, arXiv:0909.3603v1 (2009).

[110] R. Khasanov, M. Bendele, A. Amato, K. Conder, H. Keller, H.-H. Klauss, H. Luetkens, and E. Pomjakushina, *Evolution of Two-Gap Behavior of the Superconductor $FeSe_{1-x}$*, Phys. Rev. Lett. **104**, 087004 (2010).

[111] M. Bendele, S. Weyeneth, R. Puzniak, A. Maisuradze, E. Pomjakushina, K. Conder, V. Pomjakushin, H. Luetkens, S. Katrych, A. Wisniewski, R. Khasanov, and H. Keller, *Anisotropic superconducting properties of single-crystalline $FeSe_{0.5}Te_{0.5}$*, Phys. Rev. B **81**, 224520 (2010).

[112] A. T. Savici, Y. Fudamoto, I. M. Gat, T. Ito, M. I. Larkin, Y. J. Uemura, G. M. Luke, K. M. Kojima, Y. S. Lee, M. A. Kastner, R. J. Birgeneau, and K. Yamada, *Muon spin relaxation studies of incommensurate magnetism and superconductivity in stage-4 $La_2CuO_{4.11}$ and $La_{1.88}Sr_{0.12}CuO_4$*, Phys. Rev. B **66**, 014524 (2002).

[113] I. M. Reznik, F. G. Vagizov, and R. Troć, *Chemical bonding in the $UFe_{1-x}Ni_xAl$ alloys*, Phys. Rev. B **51**, 3013 (1995).

Bibliography

[114] P. W. Anderson, *Antiferromagnetism. Theory of Superexchange Interaction*, Phys. Rev. **79**, 350 (1950).

[115] J. B. Goodenough, *Theory of the Role of Covalence in the Perovskite-Type Manganites [La, M(II)]MnO$_3$*, Phys. Rev. **100**, 564 (1955).

[116] J. Kanamori, *Superexchange interaction and symmetry properties of electron orbitals*, Journal of Physics and Chemistry of Solids **10**, 87 (1959).

[117] M. J. Han, Q. Yin, W. E. Pickett, and S. Y. Savrasov, *Anisotropy, Itineracy, and Magnetic Frustration in High-T_c Iron Pnictides*, Phys. Rev. Lett. **102**, 107003 (2009).

[118] T. Imai, K. Ahilan, F. L. Ning, T. M. McQueen, and R. J. Cava, *Why Does Undoped FeSe Become a High-T_c Superconductor under Pressure?*, Phys. Rev. Lett. **102**, 177005 (2009).

[119] J. T. Park, D. S. Inosov, C. Niedermayer, G. L. Sun, D. Haug, N. B. Christensen, R. Dinnebier, A. V. Boris, A. J. Drew, L. Schulz, T. Shapoval, U. Wolff, V. Neu, X. Yang, C. T. Lin, B. Keimer, and V. Hinkov, *Electronic Phase Separation in the Slightly Underdoped Iron Pnictide Superconductor $Ba_{1-x}K_xFe_2As_2$*, Phys. Rev. Lett. **102**, 117006 (2009).

[120] R. Khasanov, S. Sanna, G. Prando, Z. Shermadini, M. Bendele, A. Amato, P. Carretta, R. De Renzi, J. Karpinski, S. Katrych, H. Luetkens, and N. D. Zhigadlo, *Tuning of competing magnetic and superconducting phase volumes in $LaFeAsO_{0.945}F_{0.055}$ by hydrostatic pressure*, Phys. Rev. B **84**, 100501 (2011).

[121] H. Maeter, H. Luetkens, Y. G. Pashkevich, A. Kwadrin, R. Khasanov, A. Amato, A. A. Gusev, K. V. Lamonova, D. A. Chervinskii, R. Klingeler, C. Hess, G. Behr, B. Büchner, and H.-H. Klauss, *Interplay of rare earth and iron magnetism in RFeAsO (R = La, Ce, Pr, and Sm): Muon-spin relaxation study and symmetry analysis*, Phys. Rev. B **80**, 094524 (2009).

Bibliography

[122] R. Khasanov, H. Luetkens, A. Amato, H.-H. Klauss, Z.-A. Ren, J. Yang, W. Lu, and Z.-X. Zhao, *Muon spin rotation studies of $SmFeAsO_{0.85}$ and $NdFeAsO_{0.85}$ superconductors*, Phys. Rev. B **78**, 092506 (2008).

[123] R. Khasanov, M. Bendele, A. Amato, P. Babkevich, A. T. Boothroyd, A. Cervellino, K. Conder, S. N. Gvasaliya, H. Keller, H.-H. Klauss, H. Luetkens, E. Pomjakushina, and B. Roessli, *Coexistence of incommensurate magnetism and superconductivity in $Fe_{1+y}Se_x Te_{1-x}$*, Phys. Rev. B **80**, 140511 (2009).

[124] Y. Laplace, J. Bobroff, F. Rullier-Albenque, D. Colson, and A. Forget, *Atomic coexistence of superconductivity and incommensurate magnetic order in the pnictide $Ba(Fe_{1-x}Co_x)_2 As_2$*, Phys. Rev. B **80**, 140501 (2009).

[125] Z. Shermadini, A. Krzton-Maziopa, M. Bendele, R. Khasanov, H. Luetkens, K. Conder, E. Pomjakushina, S. Weyeneth, V. Pomjakushin, O. Bossen, and A. Amato, *Coexistence of Magnetism and Superconductivity in the Iron-Based Compound $Cs_{0.8}(FeSe_{0.98})_2$*, Phys. Rev. Lett. **106**, 117602 (2011).

[126] R. Hu, K. Cho, H. Kim, H. Hodovanets, W. E. Straszheim, M. A. Tanatar, R. Prozorov, S. L. Bud'ko, and P. C. Canfield, *Anisotropic magnetism, resistivity, London penetration depth and magneto-optical imaging of superconducting $K_{0.80}Fe_{1.76}Se_2$ single crystals*, Superconductor Science and Technology **24**, 065006 (2011).

[127] B. Wei, H. Qing-Zhen, C. Gen-Fu, M. A. Green, W. Du-Ming, H. Jun-Bao, and Q. Yi-Ming, *A Novel Large Moment Antiferromagnetic Order in $K_{0.8}Fe_{1.6}Se_2$ Superconductor*, Chinese Physics Letters **28**, 086104 (2011).

[128] R. Khasanov, K. Conder, E. Pomjakushina, A. Amato, C. Baines, Z. Bukowski, J. Karpinski, S. Katrych, H.-H. Klauss, H. Luetkens,

A. Shengelaya, and N. D. Zhigadlo, *Evidence of nodeless superconductivity in $FeSe_{0.85}$ from a muon-spin-rotation study of the in-plane magnetic penetration depth*, Phys. Rev. B **78**, 220510 (2008).

[129] M. G. Vavilov, A. V. Chubukov, and A. B. Vorontsov, *Coexistence between superconducting and spin density wave states in iron-based superconductors: Ginzburg-Landau analysis*, Superconductor Science and Technology **23**, 054011 (2010).

[130] V. Cvetkovic and Z. Tesanovic, *Valley density-wave and multiband superconductivity in iron-based pnictide superconductors*, Phys. Rev. B **80**, 024512 (2009).

[131] A. Carrington and F. Manzano, *Magnetic penetration depth of MgB_2*, Physica C **385**, 205 (2003).

[132] P. K. Biswas, G. Balakrishnan, D. M. Paul, C. V. Tomy, M. R. Lees, and A. D. Hillier, *Muon-spin-spectroscopy study of the penetration depth of $FeTe_{0.5}Se_{0.5}$*, Phys. Rev. B **81**, 092510 (2010).

[133] H. Lei, R. Hu, E. S. Choi, J. B. Warren, and C. Petrovic, *Pauli-limited upper critical field of $Fe_{1+\delta}Te_{1-x}Se_x$*, Phys. Rev. B **81**, 094518 (2010).

[134] M. Fang, J. Yang, F. F. Balakirev, Y. Kohama, J. Singleton, B. Qian, Z. Q. Mao, H. Wang, and H. Q. Yuan, *Weak anisotropy of the superconducting upper critical field in $Fe_{1.11}Te_{0.6}Se_{0.4}$ single crystals*, Phys. Rev. B **81**, 020509 (2010).

[135] T. Kato, Y. Mizuguchi, H. Nakamura, T. Machida, H. Sakata, and Y. Takano, *Local density of states and superconducting gap in the iron chalcogenide superconductor $Fe_{1+\delta}Se_{1-x}Te_x$ observed by scanning tunneling spectroscopy*, Phys. Rev. B **80**, 180507 (2009).

[136] H. Kim, C. Martin, R. T. Gordon, M. A. Tanatar, J. Hu, B. Qian, Z. Q. Mao, R. Hu, C. Petrovic, N. Salovich, R. Giannetta, and R. Prozorov, *London penetration depth and superfluid density of*

single-crystalline $Fe_{1+y}(Te_{1-x}Se_x)$ and $Fe_{1+y}(Te_{1-x}S_x)$, Phys. Rev. B **81**, 180503 (2010).

[137] S. Weyeneth, R. Puzniak, U. Mosele, N. D. Zhigadlo, S. Katrych, Z. Bukowski, J. Karpinski, S. Kohout, J. Roos, and H. Keller, *Anisotropy of superconducting single crystal $SmFeAsO_{0.8}F_{0.2}$ studied by torque magnetometry*, J. Supercond. Nov. Magn. **22**, 325 (2009).

[138] S. Weyeneth, R. Puzniak, N. D. Zhigadlo, S. Katrych, Z. Bukowski, J. Karpinski, and H. Keller, *Evidence for two distinct anisotropies in the oxypnictide superconductors $SmFeAsO_{0.8}F_{0.2}$ and $NdFeAsO_{0.8}F_{0.2}$*, J. Supercond. Nov. Magn. **22**, 347 (2009).

[139] M. Bendele, P. Babkevich, S. Katrych, S. N. Gvasaliya, E. Pomjakushina, K. Conder, B. Roessli, A. T. Boothroyd, R. Khasanov, and H. Keller, *Tuning the superconducting and magnetic properties of $Fe_ySe_{0.25}Te_{0.75}$ by varying the iron content*, Phys. Rev. B **82**, 212504 (2010).

[140] Y. J. Uemura, G. M. Luke, B. J. Sternlieb, J. H. Brewer, J. F. Carolan, W. N. Hardy, R. Kadono, J. R. Kempton, R. F. Kiefl, S. R. Kreitzman, P. Mulhern, T. M. Riseman, D. L. Williams, B. X. Yang, S. Uchida, H. Takagi, J. Gopalakrishnan, A. W. Sleight, M. A. Subramanian, C. L. Chien, M. Z. Cieplak, G. Xiao, V. Y. Lee, B. W. Statt, C. E. Stronach, W. J. Kossler, and X. H. Yu, *Universal Correlations between T_c and n_s/m^* (Carrier Density over Effective Mass) in High-T_c Cuprate Superconductors*, Phys. Rev. Lett. **62**, 2317 (1989).

[141] A. Shengelaya, R. Khasanov, D. G. Eshchenko, D. Di Castro, I. M. Savić, M. S. Park, K. H. Kim, S.-I. Lee, K. A. Müller, and H. Keller, *Muon-Spin-Rotation Measurements of the Penetration Depth of the Infinite-Layer Electron-Doped $Sr_{0.9}La_{0.1}CuO_2$ Cuprate Superconductor*, Phys. Rev. Lett. **94**, 127001 (2005).

[142] R. Viennois, E. Giannini, D. van der Marel, and R. Černý, *Effect of Fe excess on structural, magnetic and superconducting properties of single-crystalline $Fe_{1+x}Te_{1-y}Se_y$*, Journal of Solid State Chemistry **183**, 769 (2010).

[143] W. Bao, Y. Qiu, Q. Huang, M. A. Green, P. Zajdel, M. R. Fitzsimmons, M. Zhernenkov, S. Chang, M. Fang, B. Qian, E. K. Vehstedt, J. Yang, H. M. Pham, L. Spinu, and Z. Q. Mao, *Tunable ($\delta\pi$, $\delta\pi$)-Type Antiferromagnetic Order in α-Fe(Te,Se) Superconductors*, Phys. Rev. Lett. **102**, 247001 (2009).

[144] S. Li, C. de la Cruz, Q. Huang, Y. Chen, J. W. Lynn, J. Hu, Y.-L. Huang, F.-C. Hsu, K.-W. Yeh, M.-K. Wu, and P. Dai, *First-order magnetic and structural phase transitions in $Fe_{1+y}Se_xTe_{1-x}$*, Phys. Rev. B **79**, 054503 (2009).

[145] H. Kamerlingh Onnes and W. Tuyn, *Measurements Concerning the Electrical Resistance or Ordinary Lead and of Uranium Lead Below 14K*, Comm. Phys. Lab. Leiden **160b**, 13 (1922).

[146] K. Prassides, J. Tomkinson, C. Christides, M. J. Rosseinsky, D. W. Murphy, and R. C. Haddon, *Vibrational spectroscopy of superconducting K_3C_{60} by inelastic neutron scattering*, NAT **354**, 462 (1991).

[147] A. P. Ramirez, A. R. Kortan, M. J. Rosseinsky, S. J. Duclos, A. M. Mujsce, R. C. Haddon, D. W. Murphy, A. V. Makhija, S. M. Zahurak, and K. B. Lyons, *Isotope effect in superconducting Rb_3C_{60}*, Phys. Rev. Lett. **68**, 1058 (1992).

[148] S. L. Bud'ko, G. Lapertot, C. Petrovic, C. E. Cunningham, N. Anderson, and P. C. Canfield, *Boron Isotope Effect in Superconducting MgB_2*, Phys. Rev. Lett. **86**, 1877 (2001).

[149] D. G. Hinks, H. Claus, and J. D. Jorgensen, *The complex nature of superconductivity in MgB_2 as revealed by the reduced total isotope effect*, NAT **411**, 457 (2001).

Bibliography

[150] D. G. Hinks and J. D. Jorgensen, *The isotope effect and phonons in MgB_2*, Physica C **385**, 98 (2003).

[151] D. D. Castro, M. Angst, D. G. Eshchenko, R. Khasanov, J. Roos, I. M. Savić, A. Shengelaya, S. L. Bud'ko, P. C. Canfield, K. Conder, J. Karpinski, S. M. Kazakov, R. A. Ribeiro, and H. Keller, *Absence of a boron isotope effect in the magnetic penetration depth of* MgB_2, Phys. Rev. B **70**, 014519 (2004).

[152] G.-M. Zhao, K. Ghosh, and R. L. Greene, *Colossal oxygen isotope shift of the charge-ordering transition in Nd 0.5 Sr 0.5 MnO 3*, J. Phys.: Condens. Matter **10**, L737 (1998).

[153] G. meng Zhao, H. Keller, and K. Conder, *Unconventional isotope effects in the high-temperature cuprate superconductors*, J. Phys.: Condens. Matter **13**, R569 (2001).

[154] H. Keller, *Unconventional Isotope Effects in Cuprate Superconductors*, in *Unconventional Isotope Effects in Cuprate Superconductors*, edited by K. A. Müller and A. Bussmann-Holder, Vol. 114 of *Structure & Bonding*, pages 1336–1338, Springer Berlin / Heidelberg, 2005.

[155] J. P. Franck, S. Harker, and J. H. Brewer, *Copper and oxygen isotope effects in* $La_{2-x}Sr_xCuO_4$, Phys. Rev. Lett. **71**, 283 (1993).

[156] R. Khasanov, A. Shengelaya, D. Di Castro, E. Morenzoni, A. Maisuradze, I. M. Savić, K. Conder, E. Pomjakushina, A. Bussmann-Holder, and H. Keller, *Oxygen Isotope Effects on the Superconducting Transition and Magnetic States Within the Phase Diagram of* $Y_{1-x}Pr_xBa_2Cu_3O_{7-\delta}$, Phys. Rev. Lett. **101**, 077001 (2008).

[157] M. Rotter, M. Pangerl, M. Tegel, and D. Johrendt, *Superconductivity and crystal structures of* $Ba_{1-x}K_xFe_2As_2$ *($x = 0 - 1$)*, Angew. Chem. Int. Ed. **47**, 7949 (2008).

Bibliography

[158] A. Bussmann-Holder, A. Simon, H. Keller, and A. Bishop, *Identifying the Pairing Mechanism in Fe-As Based Superconductors: Gaps and Isotope Effects*, Journal of Superconductivity and Novel Magnetism **24**, 1099 (2011), 10.1007/s10948-010-0864-z.

Bibliography

i want morebooks!

Buy your books fast and straightforward online - at one of world's fastest growing online book stores! Environmentally sound due to Print-on-Demand technologies.

Buy your books online at
www.get-morebooks.com

Kaufen Sie Ihre Bücher schnell und unkompliziert online – auf einer der am schnellsten wachsenden Buchhandelsplattformen weltweit! Dank Print-On-Demand umwelt- und ressourcenschonend produziert.

Bücher schneller online kaufen
www.morebooks.de

VDM Verlagsservicegesellschaft mbH
Heinrich-Böcking-Str. 6-8   Telefon: +49 681 3720 174   info@vdm-vsg.de
D - 66121 Saarbrücken   Telefax: +49 681 3720 1749   www.vdm-vsg.de

Printed by Books on Demand GmbH, Norderstedt / Germany